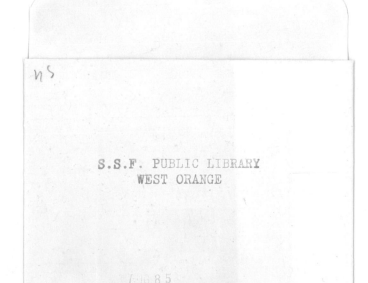

Introduction to Blueprint Reading and Sketching

Dr. C. Thomas Olivo
Albert V. Payne
Thomas P. Olivo

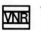 **VAN NOSTRAND REINHOLD COMPANY**
NEW YORK   CINCINNATI   TORONTO   LONDON   MELBOURNE

Printed in the United States of America
Designed by Graphics International
Albany, New York 12205

Published by Van Nostrand Reinhold Company Inc.
135 West 50th Street, New York, NY 10020

Fleet Publishers
1410 Birchmount Road
Scarborough, Ontario M1P 2E7, Canada

Van Nostrand Reinhold Australia Pty. Ltd.
480 Latrobe Street
Melbourne, Victoria 3000, Australia

Van Nostrand Reinhold Company Limited
Molly Millars Lane
Wokingham, Berkshire, England RG11 2PY

16 15 14 13 12 11 10 9 8 7 6 5 4 3 2 1

Library of Congress Cataloging in Publication Data

Olivo, C. Thomas.
Introduction to blueprint reading and sketching.

Rev. ed. of: Basic blueprint reading and sketching.
1978.
Includes index.
1. Blue-prints. 2. Freehand technical sketching.
I. Payne, Albert V. II. Olivo, Thomas P. III. Title.
T379.04 1983   604.2'5   82-20216
ISBN 0-442-27067-4

PREFACE

Blueprints are the guideposts of industry. Both simple and complex parts and mechanisms can be described graphically on blueprints with such completeness that one part, or any number of the same part, can be manufactured to the specified size, shape, and degree of accuracy. The blueprint points the way at each stage of production from the time an idea is conceived until the product is completed.

If it were possible to accumulate all of the blueprints produced over a long period of years, they would furnish an accurate inventory of manufactured articles and the evolution of industry. These blueprints would represent a partial historical record of the achievements and progress of civilization.

Blueprints tell another story, too, about international dependence and cooperation, since the blueprint is the universal language of industry. While systems of measurements and drafting techniques vary among different countries, there is an accelerating movement toward standardization. The International System of Units (SI), established by agreement among nations, provides the interlocking framework for greater acceptance of a unified metric system of measurement. The International Organization for Standardization (ISO) is also active in trying to establish standards that are acceptable throughout the world. During the many years of transition incorporating features of the customary inch and SI metric systems, workers will be required to interpret drawings in both systems and to produce identical parts and mechanisms even though they work in different lands. The universal language of the blueprint forms the foundation for the production of interchangeable parts.

Because techniques of representation differ with manufacturers, it was necessary to make careful analyses of drafting room practices and occupational studies to determine the blueprint reading needs of skilled and semiskilled workers. These comparative studies of training needs provided direction and emphasized the degree of thoroughness with which each unit of this book had to be organized and written.

Recommended standards of the American National Standards Institute (ANSI) and the International Organization for Standardization (ISO) are used in this edition of *Introduction to Blueprint Reading and Sketching*. The scope of this book insures complete mastery of blueprint reading by the reader. The units cover the range of drawings, sketches, and prints which technicians and mechanics are normally required to read and interpret accurately according to ANSI, SI Metric and other industry practices. The individual units incorporate many of the newest, tested teaching-learning methods and curriculum development practices which the writers have used successfully under actual conditions.

Grateful acknowledgment is made to Elmer A. Rotmans, former Head of the Drafting Department, Edison Technical and Industrial High School, Rochester, New York, for the skillful preparation of the drawings; and to Peter J. Olivo, former Associate Director of Curriculum Research, Delmar Publishers.

GUIDELINES FOR STUDY

Introduction to Blueprint Reading and Sketching is the basic guide for a series of trades and technical occupations clusters. This book provides instructional material and practical applications for students, apprentices, technicians, and others who must develop the ability to read and accurately interpret blueprints and to make simple sketches.

Like all former editions of *Introduction to Blueprint Reading and Sketching,* this new edition has been updated consistent with evolving standards and changing industrial practices. This format still includes the use of two colors throughout the content to help you understand and apply each new principle or concept as it is presented.

The major changes cover the following:

- Beginning with Unit 11, a number of drawings which previously were dimensioned with fractional dimensions have been changed to the two-place decimal system. On some of these drawings, nominal diameters are still given as fractional inch dimensions for drills, counterbores, taps, and reamers where such sizes represent industrial practices. Two-place decimal dimensioning provides still another variation in dimensioning techniques as experienced in shops and laboratories.

- Considerable changes were made in Unit 18 on threads and thread conventions to incorporate the latest practices of representing, dimensioning, and interpreting these features.

- Applications of latest ANSI symbols for surface finish appear in Unit 20.

- Changes were made in dimensioning with shop notes (Unit 21), particularly as these relate to knurling, grooving, chamfering, and keys and keyway cutting.

- SI metrics and the interpretation of metric drawings are now introduced in Section 4, Unit 22. The depth of coverage of metrics follows the need within industry for such drawing interpretation. Dual dimensioning (using customary inch and SI metric units) provides examples of industrial drawings of parts which require machining and measuring in either one or both systems.

- Unit 25 incorporates current techniques of sectioning.

- A new Section 6, Unit 27 has been added to cover basic symbols, representation methods, and dimensioning as applied to welding drawings. Welding principles and representation techniques are presented in order to meet the increasing demand to interpret drawings of parts fabricated by fusion processes.

- Part 2 dealing with sketching techniques has also been changed. Functional two-place decimal dimensioning is used with many of the drawings. The changes permit a great variety of dimensioning applications which cut across fractional-inch, decimal-inch, and metric measurements.

Introduction to Blueprint Reading and Sketching is divided into two major parts. Part 1 covers the basic principles of blueprint reading and provides you with applications of each new principle. Part 2 deals with the techniques of making shop sketches without the use of instruments.

In each part, the units are grouped in sections. Part 1 begins with an Introduction to Blueprint Reading followed by Section 1, Lines; Section 2, Views; Section 3, Dimensions and Notes; Section 4, The SI Metric System; Section 5, Sections; Section 6, Welding Drawings; and Section 7, Numerical Control Drawing Fundamentals. Part 2 contains Section 8, Sketching Lines and Basic Forms; Section 9, Freehand Lettering; and Section 10, Shop Sketching: Pictorial Drawings.

The units in both parts are arranged in a natural sequence of learning difficulty. Each unit includes a Basic Principle, a Blueprint, and an Assignment.

Basic Principles Series

In the Basic Principle Series, drafting principles and concepts and all the related technical information necessary to interpret each new item on a drawing are described in detail. Terminology, shop, and laboratory practices are defined and applied in operational notes of the type normally appearing on drawings. The topics in this series are arranged in the logical order of dependence of one basic principle on the next.

Blueprint Series

Each new basic principle is applied on one or more industrial blueprints. Certain changes have been made from the original drawings to provide a wide range of experiences through the use of different drafting techniques. Many of the shop and laboratory drawings for the beginning units have been simplified by removing the title blocks. The drawings are prepared according to accepted standards established by the American National Standards Institute (ANSI). Where applicable, SI metrics are used.

Assignment Series

In the Assignment Series, questions and problems are provided to cover the application of the basic principles on the drawings or blueprints for each unit. Sufficient repetition is provided to insure the mastery of each new basic principle. Typical shop terminology is used in order to develop and test your ability to read shop drawings and blueprints for required dimensions, shape descriptions, machining operations, and other essential data. This is the kind of information needed by the technician or craftsperson for the layout, fabrication, construction, assembly, testing, or operation of a single part or a mechanism of many parts.

Encircled letters and numbers are used throughout the Blueprint and Assignment Series to simplify the problems which otherwise would require lengthy, time-consuming descriptions. A ruled space is included on each assignment sheet for solutions to all problem material.

The organization of the units in Sections 8, 9, and 10 of Part 2 is identical to the organization of the units in Part 1. The first units in Section 8 cover the principles of sketching simple straight and slant lines, curved lines, circles, and rounded corners and edges. These different shapes of lines are combined in sketching more complicated parts.

The methods of straight lettering and slant lettering, with either the right or left hand, are covered in the two units in Section 9 as there is daily need for lettering and dimensioning. The five units in Section 10 deal with the techniques of making orthographic, oblique, isometric and perspective sketches, and pictorial drawings with dimensioning.

Within each unit, a few practical, simplified techniques of sketching are described and illustrated. These are followed by a blueprint which either shows how the techniques are applied to shop sketches or furnishes information from which you can make sketches, without the use of drafting instruments or equipment.

CONTENTS

Preface iv

Guidelines for Study v

UNIT 1   Bases for Interpreting Blueprints and Sketches 1

PART 1

Section 1   Lines

	Basic Principle Series		Blueprint Series		Assignment Series	
UNIT 2	The Alphabet of Lines and Object Lines	7	Pressure Pad	8	BP-2	8
3	Hidden Lines and Center Lines	9	Cam Plate	10	BP-3	10
4	Extension Lines and Dimension Lines	11	Angle Brace	12	BP-4A . . .	12
			Corner Plate	13	BP-4B . . .	13
5	Projection Lines, Other Lines, and Line Combinations.	14	Feeder Plate	16	BP-5	17

Section 2   Views

UNIT 6	Three-View Drawings.	18	Angle Bracket	22	BP-6A . . .	23
			Cross Slide	24	BP-6B. . . .	25
7	Arrangement of Views	26	Pedestal	27	BP-7	27
8	Two-View Drawings	28	End Shaft	30	BP-8A . . .	31
			Flanged Sleeve.	32	BP-8B. . . .	33
			Cover Plate.	34	BP-8C. . . .	35
9	One-View Drawings.	36	Shoulder Pin.	37	BP-9	37
10	Auxiliary Views.	38	Shaft Support	40	BP-10. . . .	41

Section 3   Dimensions and Notes

UNIT 11	Size and Location Dimensions.	42	Die Shoe	46	BP-11. . . .	47
12	Dimensioning Cylinders, Circles and Arcs	48	Gear Arm	50	BP-12. . . .	51
13	Size Dimensions for Holes and Angles	52	Sliding Support	54	BP-13. . . .	55
14	Location Dimensions for Points, Centers, and Holes	56	Coupling	58	BP-14. . . .	59
15	Dimensioning Large Arcs and Base Line Dimensions	60	Base Plate	62	BP-15. . . .	63
16	Tolerances: Fractional and Angular Dimensions	64	Sector Punch	66	BP-16. . . .	67
17	Tolerances: Unilateral and Bilateral Tolerances, and Decimal Dimensions	68	Dovetailed Slide. . . .	70	BP-17. . . .	71
18	Representing and Dimensioning External Screw Threads	72	Spindle Shaft	76	BP-18. . . .	77
19	Representing and Specifying Internal and Left-Hand Threads.	78	Rocker Base	80	BP-19. . . .	81
20	Dimensioning Tapers and Machined Surfaces . . .	82	Offset Carrier Arm . .	86	BP-20. . . .	87
21	Dimensioning with Shop Notes	88	Cutter Adapter	92	BP-21. . . .	93

| | Basic Principle Series | | Blueprint Series | | Assignment Series | |

Section 4 The SI Metric System

UNIT 22	Metric System Dimensioning, Diametral Dimensions, and ISO Symbols	94	Positioning Shaft	98	BP-22	99
23	First-Angle Projection and Dimensioning	100	Pump Systems Mount	102	BP-23	103
24	Dual System of Screw Threads, Dimensioning, and Tolerances	104	Forged Rocker Bracket	106	BP-24	107

Section 5 Sections

UNIT 25	Cutting Planes, Full Sections, and Section Lining	108	Idler Shaft Support	110	BP-25	111
26	Half Sections, Partial Sections, and Conventional Sections	112	Cone Clutch	114	BP-26A	114
			Adapter Bracket	115	BP-26B	115

Section 6 Welding Drawings

| UNIT 27 | Symbols, Representation, and Dimensioning | 117 | Welded Rotary Plug | 122 | BP-27 | 123 |

Section 7 Numerical Control Drawing Fundamentals

| UNIT 28 | Datums, Ordinate and Tabular Dimensioning | 124 | Stacker Plate | 126 | BP-28 | 127 |

PART 2

Section 8 Sketching Lines and Basic Forms

UNIT 29	Sketching Horizontal, Vertical, and Slant Lines	130	Punch Plate	132	BP-29A	133
			Tee Slide	134	BP-29B	135
30	Sketching Curved Lines and Circles	136	Shaft	138	BP-30A	139
			Slide Block	140	BP-30B	141
			Flange	142	BP-30C	143
31	Sketching Irregular Shapes	144	Tool Block	146	BP-31	147
32	Sketching Fillets, Radii, and Rounded Corners and Edges	148	Drop Forged Strap	150	BP-32A	150
			Shaft Support	151	BP-32B	151

Section 9 Freehand Lettering

| UNIT 33 | Freehand Vertical Lettering | 152 | Straight Lettering | 155 | BP-33 | 155 |
| 34 | Freehand Inclined Lettering | 156 | Slant Lettering | 157 | BP-34 | 157 |

Section 10 Shop Sketching: Pictorial Drawings

UNIT 35	Orthographic Sketching	158	Rocker Fixture	159	BP-35A	159
			Guide Bracket	160	BP-35B	161
36	Oblique Sketching	162	V-Block	165	BP-36	165
37	Isometric Sketching	166	Clutch Guide	169	BP-37	169
38	Perspective Sketching	170	Slide Block, Fixture Base	173	BP-38	173
39	Pictorial Drawings and Dimensions	174	Bearing Housing	176	BP-39	177
			Index	179		

UNIT 1
Bases for Interpreting Blueprints and Sketches

Technical information about the shape and construction of a simple part may be conveyed from one person to another by the spoken or written word. As the addition of details makes the part or mechanism more complex, the designer, draftsperson, engineer, and mechanic must use precise methods to describe the object adequately.

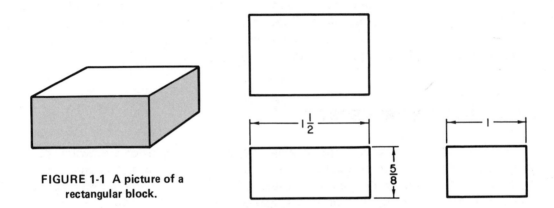

FIGURE 1-1 A picture of a rectangular block.

FIGURE 1-2 Mechanical drawing of the rectangular block.

Although a picture of a part, figure 1-1, or a photograph will help in describing an object, neither method shows the exact sizes, cutaway sections, or machining operations required. On the other hand, a drawing made accurately with instruments, figure 1-2, or a shop sketch meets the requirements for an accurate description of shape, construction and size.

BLUEPRINT READING AS A UNIVERSAL LANGUAGE

Blueprints are the universal language by which all the information the mechanic, designer, and others need to know is furnished. Blueprint reading refers to the process of interpreting a print to form a mental picture of how the object looks when completed.

Training in blueprint reading includes the development of the ability to visualize various manufacturing or fabricating processes required to make the part. Blueprint reading requires the development of certain basic principles underlying the use of various types of lines and views. The training must also include how to apply dimensions and visualizing how the inside of the part looks in section.

The mechanic must develop an understanding of the universal standards, symbols, signs, and other techniques which the draftsperson uses to describe a part, unit or mechanism completely. Technicians also develop fundamental skills in making sketches so that with pencil and paper they can record the data on the sketch relating to dimensions, notes and other details needed to construct the part.

INDUSTRIAL PRACTICES RELATING TO THE USE OF DRAWINGS

In modern industrial practice, original drawings are seldom sent to the shop or laboratory. Instead, exact reproductions of the original are made. The duplicate copies are then distributed to all individuals, departments, and plants who are responsible for planning, fabricating, or assembling a part or unit. The original drawings are filed for record purposes and for protection. Revisions and changes in design are often recorded on these drawings.

Master drawings may be made freehand, with the aid of drafting instruments, or by combining hand processes with the use of acetate sheets. These sheets are available commercially with preprinted lines, symbols and other standard parts and information. When the required item is removed from the acetate sheet, positioned and stuck to the base material, it becomes a part of the original drawing.

Many industries translate statistical and other data by using computers which feed and control automated drafting machines. These machines may be used to produce original drawings or to draw complicated shapes or circuits directly on parts that are to be machined, formed, or assembled into simple or complex units.

REPRODUCTION PROCESSES

Reproduction is the process of making one or more copies of a drawing, sketch, or other written material. A reproduction may be made to the same size as the original, or it may be an enlargement or a reduction of the original size. The reproduction may also be in a positive form like the master, or in a negative form in which light and dark areas are reversed from the master.

There are five basic reproduction processes in common use. Each process yields a particular kind of print. The kind of print depends on such requirements as time and cost factors, resistance to soiling and aging under specific work conditions, size, quantity, and desirable color.

The five basic processes may be classified as: (1) the iron process, (2) electrostatic process, (3) silver process, (4) diazo process, and (5) heat process.

The common materials used for the original drawings from which reproductions are made include paper, vellum, tracing cloth, and polyester or acetate film. The accompanying chart shows the basic reproduction processes and the materials on which drawings can be made for each process.

The Blueprint/Brownprint Iron Process

One of the oldest and still widely used practical and inexpensive processes of duplicating master drawings is called *blueprinting*. In reproducing drawings by this method, the blueprint is made on a paper or cloth coated with an emulsion which is sensitive to light. A transparent tracing of the required drawing is placed on the sensitized paper or cloth. When exposed to a strong light which passes through the transparent tracing, the lines on the drawing hold back some of the light and leave their impression on the blueprint paper. After exposure to a light source, the sensitized paper or cloth is passed through a developer, washed with water, fixed by an oxidizing agent, rewashed, and dried. The blueprint, except for slight shrinkage during the drying process, is an exact duplicate, in negative form, of the original drawing.

Basic Reproduction Processes		Basic Materials of Original Drawings		
		Paper/ Vellum	Tracing Cloth	Film: Polyester/ Acetate
A	Iron Process — Blueprint/Brownprint	X	X	
B	Electrostatic Process — Xeroxing	X		
C	Silver Process — Photocopying	X	X	X
D	Diazo Process — Whiteprint/Sepia Print	X	X	X
E	Heat Process — Blueprint/Blackprint	X		

For example, on the original drawing of the Guide Pin, the lines and dimensions are black or opaque on a transparent background material. The print of the Guide Pin, figure 1-3, is the negative of the original; that is, the lines are white on a color background.

FIGURE 1-3 Sample blueprint.

It should also be noted that figure 1-3 includes a *title block* in the lower, right-hand corner. A title block may contain such information as:

- Name of the part
- Quantity needed
- Order number
- Date
- Number of the drawing
- Scale size used
- Name of the draftsperson
- Name of drawing inspector
- Material

The term *blueprint* was originally meant to refer to a blue and white reproduction. Today it is used loosely in modern industrial language to cover many different kinds of reproductions. *Brownprints* are produced by a similar process using a coating that, after exposure to light and washing in a hypo bath, causes the paper to turn a deep brown-black. The brownprint copy is translucent and may be used as a negative master from which the same or other kinds of prints may be made.

The Electrostatic Process

In the electrostatic process a base material of paper or metal, coated with zinc oxide, is given an electrostatic charge. The original to be reproduced is either projected or brought into contact with the coating. Ultraviolet light is used. The electrostatic charge is eliminated in all areas except those which are shadowed by the markings on the original. As a result, the base material is left with an invisible electrostatic image of the original.

An electrostatically charged powder, with a charge opposite to the zinc oxide, is dusted over the base material. The powder particles adhere to the areas that still contain charged zinc oxide. When the original base material is translucent, the print is made permanent by heat fusion. Otherwise, the image of the original must be transferred to a translucent or offset plate.

The electrostatic process can be used also to produce *mats.* The mats serve as plates for offset printing. This method of reproducing large quantities of drawings and other technical data is fast and economical.

The portable *telecopier* is another application of the electrostatic process. The telecopier makes volume prints from original drawings or microfilm. The telecopier can also receive or send design, engineering, production, and other technical information contained on prints over standard telephone lines. Once a telephone circuit is established, the receivers are placed in position on the telecopier. The document is fed into the transmitting machine. The receiving machine converts the input signals into an exact reproduction of the original copy.

The Silver Process

The silver process is the oldest reproduction process and is better recognized under the names of *Photostat®, photocopy,* and *microfilm.* In each case, the reproduction is made using photographic processes and materials.

Permanently enlarged or reduced reproductions of drawings, or any other master copy, may be made on any base material that can be coated with a sensitized solution, photographed, washed, fixed in another solution for permanence, and dried.

Since there are many variations of the silver process, it is possible to produce a positive print from a positive original, a negative print from a positive, or prints using a number of other combinations. The sensitized materials may be exposed by direct contact with the master, by having a light pass through the original, or by bouncing light from the original back to the sensitized coating.

A *microfilm* is a photographically reduced copy. The microfilm is used internationally to record drawings and other significant information at a fraction of the actual size. The original copy is reduced up to 60 times. The photographed image appears on a roll of film or it may be placed individually in an *aperture card.*

Microfilm systems are built around four basic sizes: 16 mm, 35 mm, 70 mm, and 105 mm. The microfilm is read on a special enlarging *reader* for rapid viewing and legibility. Enlargements can be made electrostatically, photographically, or by photocopying processes. Reproduction of prints from microfilm is rapid. Copies may be produced automatically and in a variety of sizes.

There are many advantages to a microfilming system. There is a limited amount of space required for the storage of drawings, prints, and other data. Microfilming also helps to preserve original drawings and provides for easy reference and location. The system can easily be locked into a computer bank.

Since the *aperture card* is essentially a standard data card, specific information can be punched into it. Thus, it is possible to sort, file, and retrieve aperture cards according to the specific information required. Duplicate cards and/or reproductions of a desired drawing may then be produced.

The Diazo (Whiteprint) Process

Positive-reading reproductions with dark lines of almost any color on a white background can be produced by the diazo (Whiteprint) process. In the *dry diazo process,* the dark areas of a drawing prevent a sensitized material from being exposed to an ultraviolet light. Exposure of the print material to an ammonia atmosphere develops the color in the unexposed area. *Sepia* (yellow-brown) prints or prints with other colors may be made by varying the azo dye composition.

Wet whiteprints or sepia prints are similar to the dry prints except that a wet solution is used after exposure and the reproductions require drying. Reproductions by the diazo process are not as permanent as those resulting from other processes. The print may fade over a period of time.

The color of the lines and background on a *blue-line print* are the reverse of a blueprint. Blue-line prints are produced from a negative copy (called a *vandyke*) of the original drawing. The vandyke (negative copy) is made in the same way as a blueprint. The difference is that the background is dark brown. The lines are white. When a blueprint is made from a vandyke print, another print is produced which has blue lines against a white background.

The Heat Process

The heat process produces changes in heat-sensitive chemicals that are used to coat a base material. Dark areas on a drawing permit more heat to be absorbed by the sensitized coating than do unmarked areas. As a result, the additional heat absorbed by the chemically-treated base material produces a blue-black reproduction which is a positive print of the original.

Regardless of the method of duplicating an original drawing, the principles of graphically describing an object to furnish exact and positive information remain the same. Again, blueprint reading refers to the process of interpreting a drawing by visualizing the object clearly, accurately, and with speed and skill from the information which is presented in graphic form.

ELEMENTS COMMON TO ALL DRAWINGS

A mechanical drawing includes lines of different lengths, shapes, and intensity. These lines, when combined with each other, define the form, details, and size of an object. The lines in turn are grouped according to the surfaces of the object being viewed in what are called *views.* As a further help in describing the object, certain dimensions and notes are included on each view. Sometimes objects are cut apart by imaginary cutting planes to show a *sectional view* of the internal shape and construction of the part.

The fundamental principles of blueprint reading by which typical industrial drawings can be interpreted are included in the instructional units in the following parts and sections.

Part 1 • Section 1 . . . . .Lines
        • Section 2 . . . . .Views
        • Section 3 . . . . .Dimensions and Notes
        • Section 4 . . . . .SI Metric System
        • Section 5 . . . . .Sections
        • Section 6 . . . . .Welding Drawings
        • Section 7 . . . . .Numerical Control Drawing Fundamentals
Part 2 • Section 8 . . . . .Sketching Lines and Basic Forms
        • Section 9 . . . . .Freehand Lettering
        • Section 10 . . . . .Shop Sketching: Pictorial Drawings

PART 1

SECTION 1
Lines

UNIT 2
The Alphabet of Lines
and Object Lines

The line is the basis of all industrial drawings. By combining lines of different thicknesses, types, and lengths, it is possible to describe graphically any object in sufficient detail so that a craftsperson with a basic understanding of blueprint reading can accurately visualize the shape of the part.

THE ALPHABET OF LINES

The American National Standards Institute (ANSI) has adopted and recommended certain drafting techniques and standards for lines. The types of lines commonly found on drawings are known as the *alphabet of lines.* The six types of lines which are most widely used from this alphabet include: (1) object lines, (2) hidden lines, (3) center lines, (4) extension lines, (5) dimension lines, and (6) projection lines. A brief description and examples of each of the six types of lines are given in this section. These lines are used in combination with each other on all the prints in the Blueprint Series. Problem material on the identification of lines is included in the Assignment Series. Other lines, such as those used for showing internal part details, or specifying materials are covered in later units.

The thickness (weights) of lines are relative because they depend largely on the size of the drawing and the complexity of each member of an object. For this reason, comparative thicknesses (weights) of lines are used.

OBJECT LINES

The shape of an object is described on a drawing by thick (heavy) lines known as *visible edge* or *object lines.* An object line, figure 2-1, is always drawn thick (heavy) and solid so that the outline or shape of the object is clearly emphasized on the drawing, figure 2-2.

(SOLID, THICK LINE)

THIS TYPE OF LINE REPRESENTS THE OUTLINE OF AN OBJECT. THICKNESS OF THE LINE MAY VARY ACCORDING TO THE SIZE AND COMPLEXITY OF THE PART BEING DESCRIBED.

FIGURE 2-1 Object line.

OBJECT LINES DESCRIBE THE SHAPE OF AN OBJECT

OBJECT LINES

FIGURE 2-2 Application of the object line.

PRESSURE PAD	BP-2
QTY. 12	MAT'L. C.D.S.

VIEW I

VIEW I

VIEW II

VIEW III

VIEW II

VIEW III

PRESSURE PAD (BP-2)

ASSIGNMENT UNIT 2

Student's Name_____

1. Give the name of the part.

2. What is the number of the blueprint?

3. How many Pressure Pads are needed?

4. What name is given to the thick (heavy) line which shows the shape of the part?

5. What lettered lines show the shape of the part in:

    a. View I

    b. View II

    c. View III

1. _____

2. _____

3. _____

4. _____

5. View I _____

   View II _____

   View III _____

UNIT 3
Hidden Lines and Center Lines

HIDDEN LINES

To be complete, a drawing must include lines which represent all the edges and intersections of surfaces in the object. Many of these lines are invisible to the observer because they are covered by other portions of the object. To show that a line is hidden, the draftsperson usually uses a thin, broken line of short dashes, figure 3-1. Figure 3-2 illustrates the use of hidden lines.

(THIN, BROKEN LINE, SHORT DASHES) THIS TYPE OF LINE REPRESENTS INVISIBLE EDGES AND SURFACES

FIGURE 3-1 Hidden line.

SHOWS INVISIBLE EDGE

FIGURE 3-2 Application of hidden lines.

CENTER LINES

A center line, figure 3-3, is drawn as a thin (light), broken line of long and short dashes, spaced alternately. Center lines are used to indicate the center of a whole circle or a part of a circle, and also to show that an object is symmetrical about a line, figure 3-4.

(THIN, BROKEN LINE, LONG AND SHORT DASHES) THIS TYPE OF LINE INDICATES CENTERS OF CIRCLES, ARCS, AND SYMMETRICAL OBJECTS

FIGURE 3-3 Center line.

(A) WHOLE CIRCLES   (B) PARTS OF A CIRCLE   (C) SYMMETRICAL OBJECTS

FIGURE 3-4 Application of center lines.

REQ'D. 25
MATERIAL - BRONZE
ORDER NO. 66 A 251

CAM PLATE   BP-3

ASSIGNMENT UNIT 3

CAM PLATE (BP-3)

Student's Name _____

1. How many Cam Plates are required?

2. Name the material for the parts.

3. What type of line is used to describe the shape of the part?

4. Give the letters of all the lines which show the outside shape of the part.

5. Name the kind of line which represents an invisible edge.

6. What lettered lines in View I show the slot in the Cam Plate?

7. What line in View II shows an invisible surface?

8. What kind of line indicates the center of a circle, or part of a circle?

9. What lettered line in View I shows the circular end?

10. What lettered lines in View I are used to locate the center of the circular end?

1. _____

2. _____

3. _____

4. _____

5. _____

6. _____

7. _____

8. _____

9. _____

10. _____

UNIT 4
Extension Lines and
Dimension Lines

EXTENSION LINES

Extension lines are used in dimensioning to show the size of an object. Extension lines, figure 4-1, are thin, solid lines which extend away from an object at the exact places between which dimensions are to be placed.

FIGURE 4-1 Extension lines used
for dimensions.

FIGURE 4-2 Application of extension lines
in dimensioning.

A space of one-sixteenth inch is usually allowed between the object and the beginning of the extension line, figure 4-2.

DIMENSION LINES

Dimension lines, figure 4-3, are thin lines broken at the dimension and ending with arrowheads. The tips or points of these arrowheads indicate the exact distance referred to by a dimension placed at a break in the line.

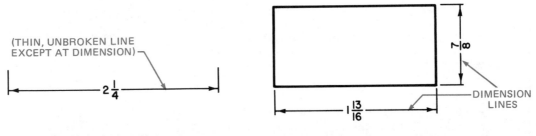

FIGURE 4-3 Dimension line.

FIGURE 4-4 Application of dimension lines.

The point or tip of the arrowhead touches the extension line. The size of the arrow is determined by the thickness of the dimension line and the size of the drawing. Closed (⟵) and open (⟶) arrowheads are the two shapes generally used. The closed arrowhead is preferred. The extension line usually projects 1/16 inch beyond a dimension line, figure 4-4. Any additional length to the extension line is of no value in dimensioning.

ANGLE BRACE

COLD DRAWN STEEL PLATE – $\frac{1}{8}$ THICK

BP-4A

ANGLE BRACE (BP-4A)

1. Name the material specified for the Angle Brace.

2. What is the overall length of each side of the Brace?

3. What is the width of each leg of the Brace?

4. What is the thickness of the metal in the Brace?

5. What is the name given to the kind of line marked Ⓐ, Ⓑ, and Ⓒ?

6. What kind of lines are Ⓓ and Ⓔ?

7. What kind of lines are Ⓕ and Ⓖ?

8. Why are object lines made thicker than extension and dimension lines?

ASSIGNMENT A UNIT 4

Student's Name _____

1. _____

2. _____

3. _____

4. _____

5. _____

6. _____

7. _____

8. _____

MALLEABLE IRON - $\frac{3}{16}$ THICK
4 REQUIRED

CORNER PLATE	BP-4B

CORNER PLATE (BP-4B)

1. What kind of line is (A)?

2. What kind of line is (B)?

3. What kind of line is (C)?

4. Determine the overall length of the Plate from left to right.

5. Determine the overall width of the Plate from top to bottom.

6. Give the center distance between the two upper holes.

7. Determine distance (D).

8. What kind of line is (E)?

9. What kind of line is (F)?

10. What radius forms the rounded corner of the Plate?

11. Name the material specified for the Plate.

12. How many Corner Plates are required?

ASSIGNMENT B UNIT 4

Student's Name _____

1. _____

2. _____

3. _____

4. _____

5. _____

6. _____

7. _____

8. _____

9. _____

10. _____

11. _____

12. _____

UNIT 5
Projection Lines, Other Lines, and Line Combinations

PROJECTION LINES

Projection lines are used by draftspersons and designers to establish the relationship of lines and surfaces in one view with corresponding points in other views. *Projection lines,* figure 5-1, are thin, unbroken lines projected from a point in one view to locate the same point in another view. Projection lines do not appear on finished drawings except where a part is complicated and it becomes necessary to show how certain details on a drawing are obtained.

FIGURE 5-1 Application of projection lines.

OTHER LINES

In addition to the six common types of lines, the alphabet of lines also includes other types such as the cutting plane line, break lines, phantom lines to indicate adjacent parts and alternate positions, and lines for repeated detail. These less frequently used lines, figure 5-2, are found in more advanced drawings, and will be described in greater detail as they are used in later drawings in this text.

(A) CUTTING PLANE LINES

(THICK) BROKEN LINE CONSISTS OF ONE LONG AND TWO SHORT DASHES, SPACED ALTERNATELY. ARROWS INDICATE THE DIRECTION IN WHICH THE SECTION IS TAKEN.

(THICK) LONG DASHES OF EQUAL LENGTH

(B) BREAK LINES

(THICK) FREEHAND LINE FOR SHORT BREAKS

(THIN) RULED LINE AND ZIG-ZIG LINES FOR LONG BREAKS

(C) PHANTOM LINE

(THIN) BROKEN LINE OF LONG DASH FOLLOWED BY TWO SHORT DASHES. INDICATES ADJACENT PARTS AND ALTERNATE POSITIONS.

FIGURE 5-2 Samples of other lines used on drawings.

LINES USED IN COMBINATION

Most drawings consist of a series of object lines, hidden lines, center lines, extension lines and dimension lines used in combination with each other to give a full description of a part or mechanism, figure 5-3.

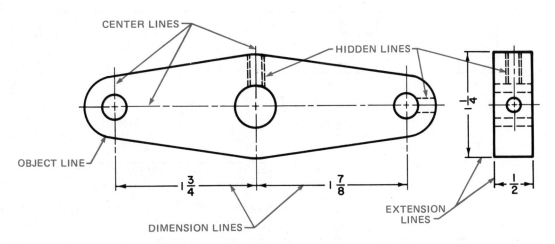

CENTER LINES

HIDDEN LINES

OBJECT LINE

$1\frac{3}{4}$

$1\frac{7}{8}$

$\frac{1}{4}$

$\frac{1}{2}$

DIMENSION LINES

EXTENSION LINES

FIGURE 5-3 Lines used in combination.

$\frac{3}{8}$ DRILL
2 HOLES

$\frac{1}{2}$ R

VIEW I

SLOT
$\frac{1}{2}$ WIDE $\frac{1}{4}$ DEEP

1 R

$3\frac{1}{2}$

$\frac{3}{4}$

$\frac{3}{4}$

$\frac{1}{2}$

VIEW II

VIEW III

VIEW I

VIEW II

VIEW III

MATERIAL: GRAY CAST IRON	
ORDER NO.	QUANTITY
2A 56 – 72	24
FEEDER PLATE	**BP-5**

FEEDER PLATE (BP-5)   **ASSIGNMENT UNIT 5**

Student's Name _____

1. What is the name of the part?   1. _____

2. What is the blueprint number?   2. _____

3. What is the Plate order number?   3. _____

4. How many parts are to be made?   4. _____

5. Name the material specified for the part.   5. _____

6. Study the Feeder Plate, BP-5.
    a. Locate and name each line from (A) to (Z) in the space provided for each in the table.
    b. Tell how each line from (A) to (L) is identified. (NOTE: Line (A) is filled in as a guide.)

Line on Drawing	(A) Name of Line	(B) How the Line is Identified
(A)	EXTENSION LINE	THIN, UNBROKEN LINE
(B)		
(C)		
(D)		
(E)		
(F)		
(G)		
(H)		
(I)		
(J)		
(K)		
(L)		

Line on Drawing	(A) Name of Line	Line	(A) Name of Line
(M)			
(N)		(U)	
(O)		(V)	
(P)		(W)	
(Q)		(X)	
(R)		(Y)	
(S)		(Z)	
(T)			

SECTION 2
Views

UNIT 6
Three-view Drawings

Regularly-shaped flat objects which require only simple machining operations are often adequately described with notes on a one-view drawing (see Unit 9). However, when the shape of the object changes, portions are cut away or relieved, or complex machining or fabricating processes must be represented on a drawing, the one view may not be sufficient to describe the part accurately.

The number and selection of views is governed by the shape or complexity of the object. A view should not be drawn unless it makes a drawing easier to read or furnishes other information needed to describe the part clearly.

Throughout this text, as the student is required to interpret more complex drawings, the basic principles underlying the use of all additional views which are needed to describe the true shape of the object will be covered. Immediate application of these principles will then be made on typical industrial blueprints.

The combination of front, top, and right side views represents the method most commonly used by draftspersons to describe simple objects. The manner in which each view is obtained and the interpretation of each view is discussed in this section.

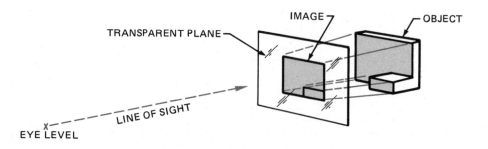

FIGURE 6-1 Projecting image for the front view.

THE FRONT VIEW

Before an object is drawn, it is examined to determine which views will best furnish the information required to manufacture the object. The surface which is to be shown as the observer looks at the object is called the *Front View*. To draw this view, the draftsperson goes through an imaginary process of raising the object to eye level and turning it so that only one side can be seen. If an imaginary transparent plane is placed between the eye and the face of the object, parallel to the object, the image projected on the plane is the same as that formed in the eye of the observer, figure 6-1.

Note in figure 6-1 that the rays converge as they approach the observer's eye. If, instead of converging, these rays are parallel as they leave the object, the image they form on the screen is equivalent to a Front View, as shown in figure 6-2.

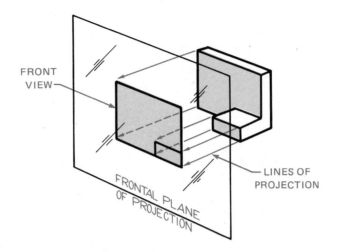

FIGURE 6-2 Front view of object.

THE TOP VIEW

To draw a *Top View,* the draftsperson goes through a process similar to that required to obtain the Front View. However, instead of looking squarely at the front of the object, the view is seen from a point directly above it, figure 6-3.

FIGURE 6-3 Projecting image to form the top view.

When the horizontal plane on which the top view is projected is rotated so that it is in a vertical plane, as shown in figure 6-4, the front and top views are in their proper relationship. In other words, the top view is always placed immediately above and in line with the front view.

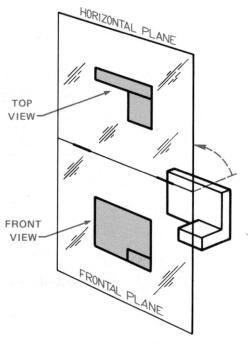

FIGURE 6-4 Relationship of front and top views.

THE SIDE VIEW

A side view is developed in much the same way that the other two views were obtained. That is, the draftsperson imagines the view of the object from the side that is to be drawn. This person then proceeds to draw the object as it would appear if parallel rays were projected upon a vertical plane, figure 6-5.

FIGURE 6-5 Projecting image to form the right side view.

FRONT, TOP, AND RIGHT SIDE VIEWS

By swinging the top of the imaginary projection box to a vertical position and the right side forward, the top view is directly above the front view, and the side view is to the right of the front view and in line with it. Figure 6-6 shows the front, top, and right side views in the positions they will occupy on a blueprint.

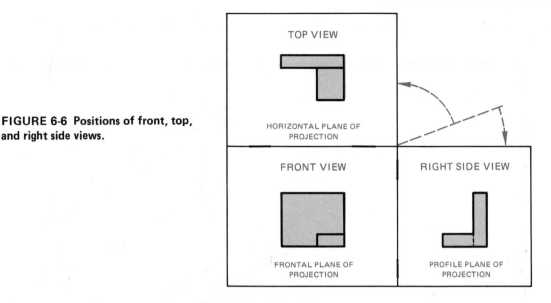

FIGURE 6-6 Positions of front, top, and right side views.

HEIGHT, WIDTH, AND DEPTH DIMENSIONS

The terms *height, width,* and *depth* refer to specific dimensions or part sizes. ANSI designations for these dimensions are shown in figure 6-7. *Height* is the vertical distance between two or more lines or surfaces (part features) which are in horizontal planes. *Width* refers to the horizontal distance between surfaces in profile planes. In the shop, the terms *length* and *width* are used interchangeably. *Depth* is the horizontal (front to back) distance between two features in frontal planes. Depth is often identified in the shop as the *thickness* of a part or feature.

HEIGHT

HORZONTAL PLANES

WIDTH

PROFILE PLANES

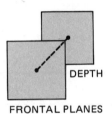

DEPTH

FRONTAL PLANES

FIGURE 6-7 Designation of height, width, and depth dimensions.

WORKING DRAWINGS

An actual drawing of a part shows only the top, front, and right side views without the imaginary transparent planes, figure 6-8. These views show the exact shape and size of the object, and define the relationship of one view to another.

A drawing, when completely dimensioned and with necessary notes added, is called a *working drawing* because it furnishes all the information required to construct the object, figure 6-9.

FIGURE 6-8 Views with transparent planes removed.

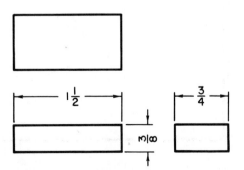

FIGURE 6-9 Example of a working drawing.

NO. REQ. 4
ORDER NO. 4 – 57
MAT. CAST IRON

ANGLE BRACKET   **BP-6A**

ANGLE BRACKET (BP-6A)

1. How many Angle Brackets are required?

2. Name the material specified for the Angle Bracket.

3. State the order number of the Bracket.

4. What is the overall width (length) of the Bracket?

5. What is the overall height?

6. What is the overall depth?

7. What is dimension (A)?

8. What is dimension (B)?

9. What surface in the top view is represented by line (C) in the right side view?

10. Name the three views that are used to describe the shape and size of the part.

11. What surface in the top view is represented by line (D) in the right side view?

12. What line in the right side view represents surface (F) in the front view?

13. What line in the right side view represents surface (J) in the front view?

14. What line in the top view represents surface (O) in the right side view?

15. What line in the front view represents surface (H) in the top view?

16. What line in the right side view represents surface (H) in the top view?

17. What kind of lines are (E) (L) (C) (D) and (K)?

18. What kinds of lines are (A) and (B)?

19. What encircled letter denotes an extension line?

20. What encircled letter in the front view denotes an object line?

ASSIGNMENT A UNIT 6

Student's Name _____

1. _____

2. _____

3. _____

4. _____

5. _____

6. _____

7. _____

8. _____

9. _____

10. _____

11. _____

12. _____

13. _____

14. _____

15. _____

16. _____

17. _____

18. _____

19. _____

20. _____

CROSS SLIDE   BP-6B

NO. REQ. 6
ORDER NO. 76 – 42
MAT. GRAY CAST IRON

ASSIGNMENT B UNIT 6

Student's Name _____

1. _____   13. _____

2. _____   14. _____

3. _____   15. _____

4. _____   16. _____

5. _____   17. _____

6. _____   18. _____

7. _____   19. _____

8. _____   20. _____

9. _____   21. _____

10. _____   22. _____

11. _____   23. _____

12. _____   24. _____

CROSS SLIDE (BP-6B)

1. What material is used for the Cross Slide?

2. How many pieces are required?

3. What is the overall width (length) of the Cross Slide?

4. What is the order number?

5. What is the overall height of the Cross Slide?

6. What are the lines marked Ⓐ and Ⓑ called?

7. What do the lines marked Ⓐ represent?

8. What two lines in the top view represent the slot shown in the front view?

9. What line in the right side view represents the slot shown in the front view?

10. What line in the front view represents surface Ⓠ in the right side view?

11. What line in the front view represents surface Ⓓ in the top view?

12. What line in the top view represents surface Ⓙ in the front view?

13. What line in the side view represents surface Ⓓ in the top view?

14. What is the diameter of the holes?

15. What is the center-to-center dimension of the holes?

16. How far is the center of the first hole from the front surface of the slide?

17. Are the holes drilled all the way through the slide?

18. What is the width of the slot shown in the front view?

19. What is the height of the slot?

20. Determine dimension Ⓢ .

21. What is the width of the projection at the top of the slide?

22. How high is the projection?

23. What kind of line is Ⓜ ?

24. What kind of line is used at Ⓞ and Ⓟ ?

UNIT 7
Arrangement of Views

The main purpose of a drawing is to give the technician sufficient information needed to build, inspect, or assemble a part or mechanism according to the specifications of the designer. Since the selection and arrangement of views depends upon the complexity of a part, only those views should be drawn which help in the interpretation of the drawing.

The average drawing which includes front, top, and side views is known as a three-view drawing. However, the designation of the views is not as important as the fact that the combination of views must give all the details of construction in the most understandable way.

The draftsperson usually selects as a front view of the object that view which best describes the general shape of the part. This front view may have no relationship to the actual front position of the part as it fits into a mechanism.

The names and positions of the different views that may be used to describe an object are illustrated in figure 7-1. Note that the back view may be placed in any one of three locations. The views which are easiest to read and, at the same time, furnish all the required information, should be the views selected for the drawing.

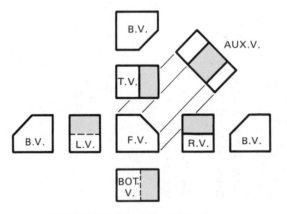

FIGURE 7-1 Identification of views.

The name and abbreviation for each view is identified throughout this text as shown in figure 7-2.

Name of View	Abbreviation
Front View	(F.V.)
Right Side View	(R.V.)
Left Side View	(L.V.)
Bottom View	(Bot. V.)
Back or Rear View	(B.V.)
Auxiliary View	(Aux. V.)
Top View	(T.V.)

FIGURE 7-2 Abbreviation of views

ASSIGNMENT UNIT 7   **PEDESTAL (BP-7)**

Student's Name _____

Place the names of the views in the spaces provided in figures 7-3, 7-4, and 7-5.

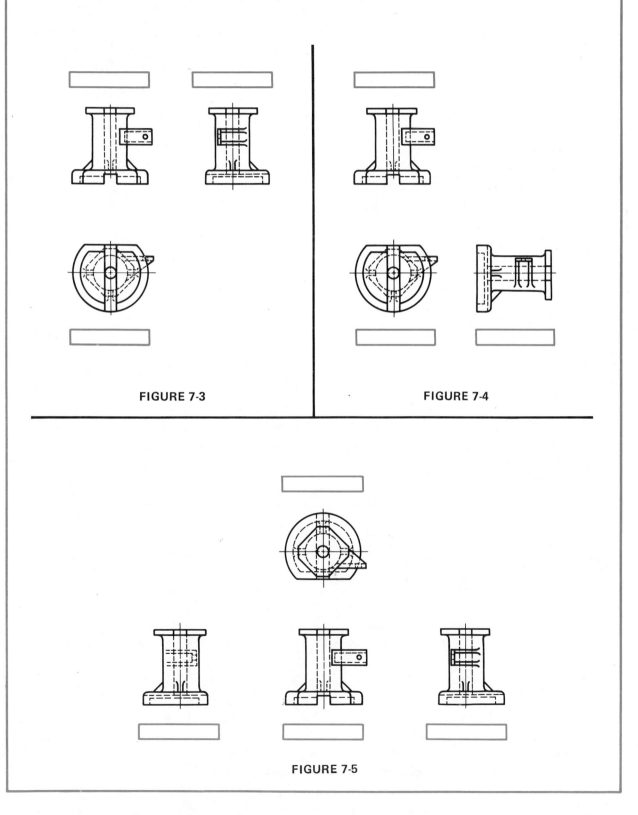

FIGURE 7-3

FIGURE 7-4

FIGURE 7-5

UNIT 8
Two-view Drawings

Simple, symmetrical flat objects and cylindrical parts, such as sleeves, shafts, rods, and studs, require only two views to give the full details of construction, figure 8-1. The two views usually include the front view and a right-side or left-side view, or a top or bottom view.

FIGURE 8-1 Examples of two-view drawings of a plug.

In the front view, figure 8-1, the center line runs through the axis of the part as a horizontal center line. If the plug is in a vertical position, the center line runs through the axis as a vertical center line.

The second view of the two-view drawing contains a horizontal and a vertical center line intersecting at the center of the circles which make up the part in this view.

The selection of views for a two-view drawing rests largely with the draftsperson or designer. Some of the combinations of views commonly used in industrial blueprints are shown in figure 8-2.

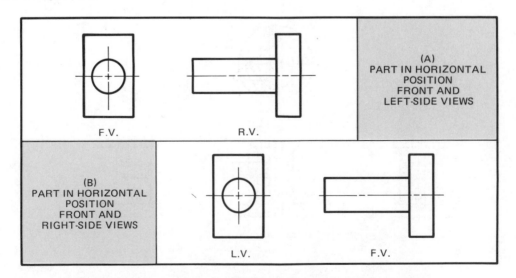

FIGURE 8-2 Combination of views for a two-view drawing of a pin.

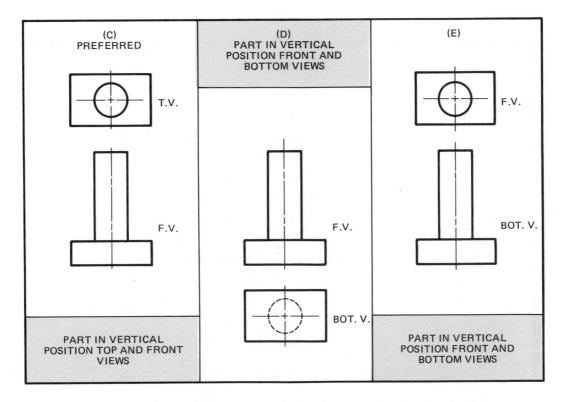

FIGURE 8-2 (Cont'd) Combination of views for a two-view drawing of a pin.

Note in figure 8-2 that different names are used to identify the same views. The name of each view depends on how the draftsperson views the object to obtain the front view.

REPRESENTING INVISIBLE CIRCLES

A hidden detail may be straight, curved, or cylindrical. Whatever the shape of the detail, and regardless of the number or positions of views, the hidden detail is represented by a hidden edge or invisible edge line, figure 8-3.

FIGURE 8-3 Use of invisible edge lines.

VIEW II

VIEW I

NO. REQ. 32	
ORDER NO. 76 – 32 – 05	
MATERIAL COLD DRAWN STEEL (C.D.S.)	
END SHAFT	**BP-8A**

END SHAFT (BP-8A)

1. Name the two views shown.

2. What line in View II represents surface (A)?

3. What lettered surface in View II represents surface (B)?

4. What circle in View II represents the 1″ hole?

5. What line in View I represents surface (H)?

6. Name line (I).

7. What kind of line is (D)?

8. Name line (J).

9. What kind of line is (K)?

10. What circle in the top view represents diameter (L)?

11. What letters in View I represent object lines?

12. What letters in Views I and II represent center lines?

13. Give the diameter of (L).

14. What is the smallest diameter of the shaft?

15. Determine the height (length) of the 1 1/4″ diameter portion.

16. What is the height (length) of the rectangular part of the shaft?

17. Give the dimensions of the rectangular part.

18. Give the overall height (length) of the shaft.

19. What is the order number?

20. State the material from which the shaft is to be machined.

ASSIGNMENT A UNIT 8

Student's Name _____

1. _____

2. _____

3. _____

4. _____

5. _____

6. _____

7. _____

8. _____

9. _____

10. _____

11. _____

12. _____

13. _____

14. _____

15. _____

16. _____

17. _____

18. _____

19. _____

20. _____

3 $\frac{1}{4}$

1 $\frac{3}{4}$

2 $\frac{1}{2}$

$\frac{7}{8}$

$\frac{1}{2}$

$\frac{1}{4}$

$\frac{1}{2}$

$\frac{5}{8}$

Ⓐ
Ⓑ
Ⓒ
Ⓓ
Ⓔ
Ⓕ
Ⓖ

NO. REQ. 12

ORDER NO. 45 – 21

MAT. C.D.S. (SAE 1020)

FLANGED SLEEVE   **BP-8B**

FLANGED SLEEVE (BP-8B)

1. What is the name of the part?

2. What is the order number?

3. How many pieces are required?

4. What material is used?

5. Name the two views which are used to represent the Flanged Sleeve.

6. Name the kind of line indicated by each of the following encircled letters.

   (A)

   (B)

   (C)

   (D)

   (E)

   (F)

   (G)

7. What is the outside diameter of both flanges?

8. What is the height (thickness) of each flange?

9. What is the diameter of the center hole?

10. Does the hole go all the way through the center of the sleeve?

11. What is the diameter of the hidden circle?

12. Determine the total or overall length of the Flanged Sleeve.

ASSIGNMENT B UNIT 8

Student's Name _____

1. _____

2. _____

3. _____

4. _____

5. _____

6.

   (A) _____

   (B) _____

   (C) _____

   (D) _____

   (E) _____

   (F) _____

   (G) _____

7. _____

8. _____

9. _____

10. _____

11. _____

12. _____

NO. REQ. 2

ORDER NO. 4 – 76

MAT'L. BRASS CASTING

COVER PLATE   **BP-8C**

COVER PLATE (BP-8C)

1. Name the material specified for the part.

2. Name the two views used to describe the part.

3. Identify the kind of line indicated by each of the following encircled letters.

   Ⓔ
   Ⓕ
   Ⓖ
   Ⓗ
   Ⓘ

4. What is the overall depth Ⓐ ?

5. What is the overall length Ⓑ ?

6. How many holes are to be drilled?

7. What is the thickness (height) of the plate?

8. What is the diameter of the holes?

9. What is the distance between the center of one of the two upper holes and the center line of the Plate?

10. Give the center distance Ⓒ of the two upper holes?

11. What is the radius that forms the two upper rounds of the Plate?

12. What radius forms the lower part of the Plate?

13. What kind of line is drawn through the center of the Plate?

14. Determine distance Ⓓ .

15. How much stock is left between the edge of one of the upper holes and the outside of the piece?

ASSIGNMENT C UNIT 8

Student's Name _____

1. _____

2. _____

3.
   Ⓔ _____
   Ⓕ _____
   Ⓖ _____
   Ⓗ _____
   Ⓘ _____

4. _____

5. _____

6. _____

7. _____

8. _____

9. _____

10. _____

11. _____

12. _____

13. _____

14. _____

15. _____

UNIT 9
One-view Drawings

Many parts which are uniform in shape require only one view to describe them adequately. This is particularly true of cylindrical objects where a one-view drawing saves drafting time and simplifies the reading of the blueprint.

When a one-view drawing of a cylinder is used, the dimensions showing the diameter must be followed by the letters *DIA* or *D* as shown in figure 9-1. This abbreviation of DIA or D for diameter and the use of a center line indicate that the part is cylindrical.

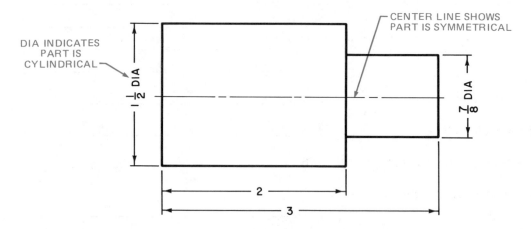

FIGURE 9-1 One-view drawing of a cylindrical shaft.

The one-view drawing is also used extensively for flat parts. With the addition of notes to supplement the dimensions on the view, the one view furnishes all the necessary information for accurately describing the part, figure 9-2.

FIGURE 9-2 One-view drawing of a flat part.

| TOOL STEEL | 245 – A – 12 | 16 | SHOULDER PIN | BP-9 |
| MATERIAL | ORDER NO. | QTY. | | |

SHOULDER PIN (BP-9)

ASSIGNMENT UNIT 9

Student's Name_____

1. Name the view represented on BP-9.

2. What is the shape of the Shoulder Pin?

3. How many outside diameters are shown?

4. What is the largest diameter?

5. What diameter is the smallest hole?

6. What is the overall length of the pin?

7. How deep is the 3/4" hole?

8. How wide is the 1.750" DIA portion?

9. What letters represent object lines?

10. What kinds of lines are Ⓑ and Ⓓ?

11. What letter represents the center line?

12. What does the center line indicate about the holes and outside diameters?

13. Give the thickness of the 2 1/4" DIA portion.

14. State the order number of the part.

15. What material is specified for the pins?

1. _____

2. _____

3. _____

4. _____

5. _____

6. _____

7. _____

8. _____

9. _____

10. _____

11. _____

12. _____

13. _____

14. _____

15. _____

UNIT 10
Auxiliary Views

As long as all the surfaces of an object are parallel or at right angles to one another, they may be represented in one or more regular views, figure 10-1.

OBJECT WITH ALL
SURFACES PARALLEL
OR AT RIGHT ANGLES
TO EACH OTHER

FIGURE 10-1 Representation of object using regular views.

The surfaces of such objects can be projected in their true sizes and shapes on either a horizontal or a vertical plane, or on any combination of these planes.

Some objects have one or more surfaces which slant and are inclined away from either a horizontal or vertical plane. In this situation the regular views will not show the true shape of the inclined surface, figure 10-2. If the true shape must be shown, the drawing must include an *auxiliary view* to represent the angular surface accurately. The auxiliary view is in addition to the regular views.

TOP VIEW

AUXILIARY VIEW
NEEDED TO SHOW THE TRUE
SHAPE OF THE INCLINED SURFACE

RIGHT SIDE VIEW

FRONT VIEW

FIGURE 10-2 Object requiring use of auxiliary view.

APPLICATION OF AUXILIARY VIEWS

Auxiliary views may be full views or partial views. Figure 10-3 shows an auxiliary view in which only the inclined surface and other required details are included. In an auxiliary view, the inclined surface is projected on an imaginary plane which is parallel to it. Rounded surfaces and circular holes, which are distorted and appear as ellipses in the regular views, will appear in their true shapes and sizes in an auxiliary view.

Auxiliary views are named according to the position from which the inclined face is seen. For example, the auxiliary view may be an auxiliary front, top, bottom, left side, or right side view. On drawings of complex parts involving compound angles, one auxiliary view may be developed from another auxiliary view. The first auxiliary view is called the *primary* view, and those views developed from it are called *secondary* auxiliary views. For the present, attention is focused on primary auxiliary views.

SUMMARY

Auxiliary views are usually partial views which show only the inclined surface of an object. In figure 10-3, the true size and shape of surface Ⓐ is shown in the auxiliary view of the angular surface.

FIGURE 10-3 Application of an auxiliary view.

The draftsperson develops this view by projecting lines ①, ②, ③, ④ and ⑤ at right angles to surface Ⓐ. When the front view is compared with the auxiliary partial view, it can be seen that the hole in surface Ⓐ appears as an ellipse in the front view. The depth of surface Ⓐ is foreshortened in the regular view while the true shape and size are shown on the auxiliary view.

In figure 10-3, the combination of *left side view, auxiliary partial top view* and *front view,* when properly dimensioned, shows all surfaces in their true size and shape. As a result, these are the only views required to describe this part completely.

.375 REAM
2 HOLES

VIEW Ⅲ

VIEW Ⅱ

VIEW Ⅰ

SHAFT SUPPORT

NO. REQ. 20

ORDER NO. 71 – 24

MAT'L. MALLEABLE CAST IRON

BP-10

ASSIGNMENT UNIT 10

Student's Name _____

SHAFT SUPPORT (BP-10)

1. Name the material specified for the part.

2. How many pieces are required?

3. What is the order number?

4. Name each of the following:

   View I

   View II

   View III

5. What kind of line is (C)?

6. What kind of line is (P)?

7. Give the diameter of hole (F).

8. Determine dimension (A).

9. Determine dimension (B).

10. What surface in the front view is the same as surface (D) in the auxiliary view?

11. How many .375'' DIA holes must be reamed in the support?

12. What is the center-to-center distance of the .375'' DIA holes?

13. What surface in the top view is represented by the line (E) in the auxiliary view?

14. What surface in the top view is represented by line (G) in the front view?

15. What line in the top view represents surface (H) in the front view?

16. What line in the front view represents surface (J) in the top view?

17. What do lines (K) and (R) represent?

18. What line in the front view represents surface (N) in the top view?

19. Determine the vertical distance between the surface represented by the line (P) and the surface represented by line (Q).

20. What is the distance between the surface represented by line (E) and the surface represented by line (S)?

1. _____

2. _____

3. _____

4. (I) _____

   (II) _____

   (III) _____

5. _____

6. _____

7. _____

8. _____

9. _____

10. _____

11. _____

12. _____

13. _____

14. _____

15. _____

16. _____

17. _____

18. _____

19. _____

20. _____

UNIT 11
Size and Location Dimensions

Drawings consist of several types of lines which are used singly or in combination with each other to describe the shape and internal construction of an object or mechanism. However, to construct or machine a part, the blueprint or drawing must include dimensions which indicate exact sizes and locations of surfaces, indentations, holes, and other details.

The lines and dimensions, in turn, are supplemented by notes which give additional information. This information includes the kind of material used, the degree of machining accuracy required, details regarding the assembly of parts, and any other data which the craftsperson needs to know to make and assemble the part.

THE LANGUAGE OF DRAFTING

To insure some measure of uniformity in industrial drawings, the American National Standards Institute (ANSI) has established drafting standards. These standards are called the language of drafting and are in general use throughout the United States. While these drafting standards or practices may vary in some respects between industries, the principles are basically the same. The practices recommended by ANSI for dimensioning and for making notes are followed in this section.

Standards for Dimensioning

All drawings should be dimensioned completely so that a minimum of computation is necessary, and the parts can be built without scaling the drawing. However, there should not be a duplication of dimensions unless such dimensions make the drawing clearer and easier to read.

Many parts cannot be drawn full size because they are too large to fit a standard drawing sheet, or too small to have all details shown clearly. The draftsperson can, however, still represent such objects either by reducing (in the case of large objects) or enlarging (for small objects) the size to which the drawing is made.

This practice does not affect any dimensions as the dimensions on a drawing give the actual sizes. If a drawing 6″ long represents a part 12″ long, a note should appear in the title box of the drawing to indicate the *scale* that is used. This scale is the ratio of the drawing size to the actual size of the object. In this case, the scale 6″ = 12″ is called *half scale* or 1/2″ = 1″. Other common scales include the one-quarter scale (1/4″ = 1″), one-eighth scale (1/8″ = 1″), and the double size scale (2″ = 1″).

CONSTRUCTION DIMENSIONS

Dimensions used in building a part are sometimes called *construction dimensions*. These dimensions serve two purposes: (1) they indicate size, and (2) they give exact locations. For example, to drill a through hole in a part, the technician must know the diameter of the hole, and the exact location of the center of the hole, figure 11-1.

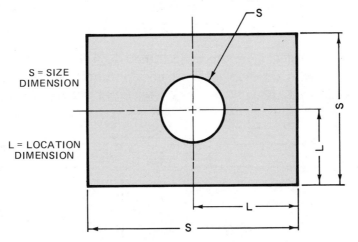

FIGURE 11-1 Dimensions indicating size and location.

TWO-PLACE DECIMAL DIMENSIONS

Dimensions may appear on drawings as *two-place decimals.* These are becoming widely used when the range of dimensional accuracy of a part is between 0.01" larger or smaller than a specified dimension (nominal size). Where possible, two-place decimal dimensions are given in even hundredths of an inch.

Three- and four-place decimal dimensions continue to be used for more precise dimensions requiring machining accuracies in thousandths or ten-thousandths of an inch.

SIZE DIMENSIONS

Every solid has three size dimensions: depth or thickness, length or width, and height. In the case of a regular prism, two of the dimensions are usually placed on the principal view and the third dimension is placed on one of the other views, figure 11-2.

FIGURE 11-2 Placing size dimensions.

LOCATION DIMENSIONS

Location dimensions are usually made from either a center line or a finished surface. This practice is followed to overcome inaccuracies due to variations in measurement caused by surface irregularities.

Draftspersons and designers must know all the manufacturing processes and operations which a part must undergo in the shop. This technical information assists them in placing size and location dimensions, required for each operation, on the drawing. Figure 11-3 shows how size and location dimensions are indicated.

FIGURE 11-3 Size and location dimensions.

PLACING DIMENSIONS

In dimensioning a drawing, the first step is to place extension lines and external center lines where needed, figure 11-4.

FIGURE 11-4 Placing extension lines
and center lines.

FIGURE 11-5 Leaders used for
dimensioning.

Dimension lines and leaders are added next. The term *leader* refers to a thin, inclined straight line terminating in an arrowhead. The leader directs attention to a dimension or note and the arrowhead identifies the feature to which the dimension or note refers. A few sample leaders are given in figure 11-5.

The common practice in placing dimensions is to keep them outside the outline of the object. The exception is where the drawing may be made clearer by inserting the dimensions within the object. When a dimension applies to two views, it should be placed between the two views as shown in figure 11-6.

FIGURE 11-6 Placing dimensions between views.

Continuous Dimensions

Sets of dimension lines and numerals should be placed on drawings close enough so they may be read easily without any possibility of confusing one dimension with another. If a series of dimensions is required, the dimensions should be placed in a line as continuous dimensions, figure 11-7A. This method is preferred over the staggering of dimensions, figure 11-7B, because of ease in reading, appearance, and simplified dimensioning.

(A) PREFERRED METHOD
CONTINUOUS DIMENSIONS

(B) NOT RECOMMENDED
STAGGERED DIMENSIONS

FIGURE 11-7 Placing series dimensions.

DIMENSIONS IN LIMITED SPACES

In the dimensioning of grooves or slots, the dimension, in many cases, extends beyond the width of the extension line. In such instances, the dimension is placed on either side of the extension line, or a leader is used, figure 11-8 (A, B, C, and D). Dots are also used on drawings for dimensioning when space is limited as shown in figure 11-8E.

FIGURE 11-8 Dimensioning grooves and slots in limited spaces.

DIE SHOE

BP-11

MATERIAL:
GRAY CAST IRON

QTY.
REQ'D.   10

ORDER NO. 10 – 51 – 17A

PART NO. CASTING 2L 715

ASSIGNMENT UNIT 11

Student's Name _____

NOTE: STUDY THE SKETCH OF THE
DIE SHOE

1. Give the dimensions in both views in the
circles provided.

2. Indicate one letter on the two-view drawing which identifies each of the following
types of lines:

(a) Extension _____

(b) Dimension _____

(c) Object _____

(d) Center line _____

(e) Hidden edge _____

3. Place an X in the correct block to identify
the kind of dimension each letter refers
to.

Dimension	Size	Location
	Dimension	
G is a		
H is a		
I is a		

DIE SHOE (BP-11)

UNIT 12
Dimensioning Cylinders, Circles, and Arcs

ALIGNED AND UNIDIRECTIONAL METHODS OF PLACING AND READING DIMENSIONS

There are two standard methods of placing dimensions. In the older method, called the *aligned* method, each dimension is placed in line with the dimension to which it refers, figure 12-1A. The second method, the *unidirectional* method, has all numbers or values placed horizontally (one direction), regardless of the direction of the dimension line. All values are read from the bottom, figure 12-1B.

FIGURE 12-1 Aligned and unidirectional dimensioning.

The aligned and unidirectional methods are illustrated in figures 12-1A and 12-1B on similar drawings of the same part. Note at (A) that the aligned dimensions are read from both the bottom and the right side. By contrast, the unidirectional dimensions, figure 12-1B, are read from the bottom (one direction only).

DIMENSIONING CYLINDERS

The length and diameter of a cylinder are usually placed in the view which shows the cylinder as a rectangle, figure 12-2. This method of dimensioning is preferred because on small diameter cylinders and holes a dimension placed in the hole is confusing.

FIGURE 12-2 Dimensioning cylinders.

Many round parts, with cylindrical surfaces symmetrical about the axis, can be represented on one-view drawings. The abbreviation for diameter, DIA, is used with the dimension in such instances because no other view is needed to show the shape of the surface, figure 12-3A. A diameter may also be identified according to ANSI standards by using the symbol Ø preceding the dimension. In shop practice, the symbol Ø often appears after the dimension. On two-view drawings, DIA may be omitted, figure 12-3B. In other words, when a cylinder is dimensioned, DIA should follow the dimension unless it is evident that the dimension refers to a diameter.

(A) ONE-VIEW DRAWING   (B) TWO-VIEW DRAWING

FIGURE 12-3 Dimensioning circles.

DIMENSIONING ARCS

An arc is always dimensioned by giving the radius. The dimension should be followed by an R, and should have an arrowhead only at the outer end, figure 12-4.

FIGURE 12-4 Dimensioning arcs.

FILLETS AND ROUNDS $\frac{1}{8}$ R

F (.257) DRILL

$\frac{5}{8}$ DIA

MATERIAL:	QTY.	250
S.A.E. 1040	REQ'D.	
ORDER NO. 16 – 37L		
PART NO. FORGING 37L		

GEAR ARM.

BP-12

ASSIGNMENT UNIT 12

Student's Name _____

NOTE: STUDY THE SKETCH OF THE
        GEAR ARM

1. Give the dimensions required in both
   views in the circles provided, using the
   unidirectional method.

2. Dimension arcs Ⓐ and Ⓑ.

3. What outside diameter is the upright
   portion? _____

4. Give the overall length of the elongated slot. _____

5. Determine the overall length of the
   Gear Arm. _____

6. Name the two views.
   _____ and _____ views.

7. Refer to the two-view drawing. Give
   one letter which identifies each type of
   line.

   (a) Center line _____
   (b) Object line _____
   (c) Extension line _____
   (d) Hidden edge _____
   (e) Dimension line _____

GEAR ARM (BP-12)

DRILL

ARC Ⓐ

Ⓔ

Ⓑ R

Ⓒ

Ⓓ

FILLETS AND ROUNDS $\frac{1}{8}$R

Ⓕ

Ⓖ

Ⓗ

Ⓘ

Ⓙ

UNIT 13
Size Dimensions for Holes and Angles

DIMENSIONING HOLES

The diameters of holes which are to be formed by drilling, reaming, or punching should have the diameter, preferably on a leader, followed by a note indicating ① the operation to be performed, and ② the number of holes to be produced, figure 13-1.

FIGURE 13-1 Dimensioning holes.

DIMENSIONING COUNTERBORED HOLES

A *counterbored hole,* figure 13-2, is one that has been machined to a larger diameter for a specified depth so that a bolt or pin will fit into this recessed hole. The counterbored hole provides a flat surface for the bolt or pin to seat against.

FIGURE 13-2 Dimensioning counterbored holes.

Counterbored holes are dimensioned by giving ① the diameter of the drill, ② the diameter of the counterbore, ③ the depth, and ④ the number of holes, figure 13-2.

DIMENSIONING COUNTERSUNK HOLES

A *countersunk hole,* figure 13-3, is a cone-shaped recess machined in a part to receive a cone-shaped flat head screw or bolt.

FIGURE 13-3 Dimensioning countersunk holes.

Countersunk holes are dimensioned by giving ① the diameter of the hole, ② the angle at which the hole is to be countersunk, ③ the diameter at the large end of the hole, and ④ the number of holes to be countersunk.

DIMENSIONING ANGLES

The design of a part may require some lines to be drawn at an angle. The amount of the divergence (the amount the lines move away from each other) is indicated by an angle measured in degrees or fractional parts of a degree. The degree is indicated by the symbol ° placed after the numerical value of the angle. For example, in figure 13-4B, 45° indicates that the angle measures 45 degrees.

Two common methods of dimensioning angles show (1) linear dimensions or (2) the angular measurement as illustrated in figure 13-4.

FIGURE 13-4 Methods of dimensioning angles.

The dimension line for an angle should be an arc whose ends terminate in arrowheads. The numeral indicating the number of degrees in the angle is read in a horizontal position, except where the angle is large enough to permit the numerals to be placed along the arc, figure 13-5.

FIGURE 13-5 Placing angular dimensions.

SLIDING SUPPORT   BP-13

MAT'L. GRAY CAST IRON

ORDER NO. 13310

NO. REQ. 5

VIEW III

BOSS

$\frac{3}{16}$ DRILL
60° CSK. TO .44 DIA
4 HOLES

VIEW II

VIEW I

.500 REAM
3 HOLES

$\frac{3}{16}$ DRILL, $\frac{3}{8}$ C'BORE
.25 DEEP, 2 HOLES

DOVETAIL

SLIDING SUPPORT (BP-13)

1. Name view I which shows the shape of the dovetail.

2. Name view III in which the bottom pad appears as a circle.

3. Name view II.

4. Name the kind of line shown at (E).

5. What surface in view I is represented by line (E)?

6. Name the kind of line shown at (G).

7. What line in view III represents surfaces (G)?

8. Name the kind of line shown at (H)?

9. What line in view III represents the line (H)?

10. Name the kind of line shown at (I).

11. Name the kind of line shown at (J).

12. What lines in the top view represent the dovetail?

13. What does the line (E) in the front view represent?

14. Determine height (A).

15. How many bosses are shown on the uprights?

16. What is the outside diameter of the boss?

17. Determine dimension (B).

18. How far off from the center of the support is the center of the two holes in the bosses of the uprights?

19. Give the dimensions for the counterbored holes.

20. What dimensions are given for the countersunk holes?

21. Give the dimensions of the reamed holes.

22. What is the dimension (C)?

23. How wide is the opening in the dovetail?

24. How deep is the dovetail machined?

25. What is the angle to the horizontal at which the dovetail is cut?

ASSIGNMENT UNIT 13

Student's Name _____

1. _____
2. _____
3. _____
4. _____
5. _____
6. _____
7. _____
8. _____
9. _____
10. _____
11. _____
12. _____
13. _____
14. _____
15. _____
16. _____
17. _____
18. _____
19. _____
20. _____
21. _____
22. _____
23. _____
24. _____
25. _____

UNIT 14
Location Dimensions for Points, Centers, and Holes

DIMENSIONING A POINT OR A CENTER

A point or a center of an arc or circle is generally measured from two finished surfaces. This method of locating the center is preferred to making an angular measurement.

In figure 14-1, the center of the circle and arc may be found easily by scribing the vertical and horizontal center lines from the machined surfaces.

FIGURE 14-1 Dimensioning the center of a circle.

DIMENSIONING EQUALLY SPACED HOLES ON A CIRCLE

If a number of holes are to be equally spaced on a circle, the exact location of the first hole is given by location dimensions. To locate the remaining holes, the location dimensions are followed by ① the diameter of the holes, ② the number of holes, and ③ the notation "equally spaced," figure 14-2.

FIGURE 14-2 Dimensioning holes equally spaced on a circle.

DIMENSIONING UNEQUALLY SPACED HOLES ON A CIRCLE

When holes are to be located on a circle, the diameter of the circle should be given so as to fix the exact center of each hole. The size and position of each hole are noted on the drawing, figure 14-3A. If more than one hole is the same diameter, then a notation may be used to indicate this fact, figure 14-3B.

For example, the notation .500 REAM – 5 HOLES means that the five holes on the drawings are reamed 1/2'' in diameter.

(A) POSITION AND SIZE
OF EACH HOLE GIVEN

(B) NOTATION WHERE HOLES
ARE SAME SIZE

FIGURE 14-3 Two methods of dimensioning unequally
spaced holes on a circle.

DIMENSIONING HOLES NOT ON A CIRCLE

Holes are often dimensioned in relation to one another and to a finished surface. Dimensions are usually given, in such cases, in the view which shows the shape of the holes, that is, square, round, or elongated. The preferred method of placing these dimensions is shown in figure 14-4A.

(A)
PREFERRED

(B)
POOR
PRACTICE

FIGURE 14-4 Placing dimensions for holes.

60°

30°

K

J

2.62

N

.34 DRILL
7 HOLES
EQUALLY
SPACED

A

.375 REAM
4 HOLES

B

3.76

I

3.25

1.000 BORE

H

G

F

E

C

1.76

D

.12 R

2.00

1.12

.50

.62

1.25

4.50

COUPLING   BP-14

COUPLING (BP-14)

1. Name the view which shows the width (height) of the Coupling.

2. Name the view in which the holes are shown as circles.

3. Name the kind of line shown at (E) (F) (G) (H) (I) (J).

4. What circle represents the 4.50″ diameter flange?

5. What circle represents the 1.000″ bore diameter?

6. Name the kind of line shown at (N).

7. How many holes are to be drilled in the larger flange?

8. Indicate the drill size to be used.

9. Give the diameter circle on which the equally spaced holes are drilled in the larger flange.

10. How many holes are to be reamed in the smaller flange?

11. How deep is the 1.25″ diameter hole bored?

12. Give the diameter of the reamed holes.

13. State the angle with the horizontal center line used for locating the first clockwise reamed hole.

14. What is the overall width (height) of the Coupling?

15. What is the diameter of the smaller flange?

16. What is the diameter of the circle (A)? (B)?

17. What is the length of the 1.000″ bored hole?

18. What is the width of the larger flange?

ASSIGNMENT UNIT 14

Student's Name _____

1. _____

2. _____

3. (E) _____
   (F) _____
   (G) _____
   (H) _____
   (I) _____
   (J) _____

4. _____

5. _____

6. _____

7. _____

8. _____

9. _____

10. _____

11. _____

12. _____

13. _____

14. _____

15. _____

16. (A) _____
    (B) _____

17. _____

18. _____

19. _____

20. (C) _____
    (D) _____

19. If a 5/16″ bolt is used in the drilled holes, what will be the clearance between the hole and the bolt?

20. Determine distances (C), (D).

UNIT 15
Dimensioning Large Arcs
and Base Line Dimensioning

DIMENSIONING ARCS WITH CENTERS OUTSIDE THE DRAWING

When the center of an arc falls outside the limits of the drawing, a broken dimension line is used as illustrated in figure 15-1. This dimension line gives the size of the arc and indicates that the arc center lies on a center line outside the drawing. This technique is used, also, when a dimension line interferes with other parts of a drawing.

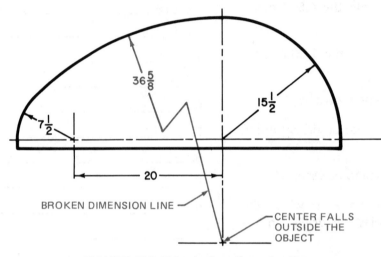

FIGURE 15-1 Using broken dimension line.

BASE LINE DIMENSIONING

In base line dimensioning, all measurements are made from common finished surfaces called *base lines* or *reference lines,* figure 15-2. Base line dimensioning is used where accurate

FIGURE 15-2 Base line dimensioning from two machined edges.

layout work to precision limits is required. Errors are not cumulative with this type of dimensioning because all measurements are taken from the base lines.

Dimensions and measurements may be taken from one or more base lines. In figure 15-2, the two base lines are at right angles to each other. The horizontal dimensions are measured from base line (B) which is a machined edge. The vertical dimensions are measured from surface (A) which is at a right angle to surface (B) and is also a machined surface.

An application of base line dimensioning, where a center line is used as the reference line, is shown in figure 15-3A. Base line dimensioning may also be applied to irregular shapes such as the template shown in figure 15-3B.

CENTER LINE USED AS BASE LINE

BASE LINE DIMENSIONING APPLIED TO IRREGULAR SHAPE

FIGURE 15-3 Applications of base line dimensioning.

Base line dimensions simplify the reading of a drawing and also permit greater accuracy in making the part.

SHOP DIMENSIONS

In most machine and metal shops, dimensions under 72 inches are usually stated in inches. By comparison, measurements for structural work and the building industry are usually given in feet and inches.

NO. REQ. 4

MAT'L. — TOOL STEEL (SAE 1090)

BASE PLATE   BP-15

BASE PLATE (BP-15)

ASSIGNMENT UNIT 15

Student's Name _____

1. Give the name of the part.

2. What material is used for the Base Plate?

3. How many parts are required?

4. What is the length of the Base Plate?

5. What is the width of the Base Plate?

6. What is the thickness of the Plate?

7. How many 5/16″ holes are to be drilled?

8. How many 3/16″ holes are to be drilled?

9. How many 1/4″ holes are to be drilled?

10. Give (a) the diameter and (b) the depth of counterbore for the 1/4″ holes.

11. What system of dimensioning is used on this drawing?

12. Give the letter of the base line in the Top View from which all depth dimensions are taken.

13. Give the letter of the base line in the Top View from which all horizontal dimensions are taken.

14. Compute the following depth dimensions: Ⓒ Ⓓ Ⓜ Ⓕ.

15. Compute the following horizontal dimensions: Ⓖ Ⓗ Ⓙ Ⓚ.

16. Compute dimensions Ⓝ and Ⓞ.

17. Give the radius to which the corners are rounded.

18. What is the radius of arc Ⓛ?

19. What letter indicates the center for arc Ⓛ?

20. Compute dimensions Ⓟ and Ⓠ.

1. _____

2. _____

3. _____

4. _____

5. _____

6. _____

7. _____

8. _____

9. _____

10. (a) _____ (b) _____

11. _____

12. _____

13. _____

14. Ⓒ = _____
    Ⓓ = _____
    Ⓜ = _____
    Ⓕ = _____

15. Ⓖ = _____
    Ⓗ = _____
    Ⓙ = _____
    Ⓚ = _____

16. Ⓝ = _____
    Ⓞ = _____

17. _____

18. _____

19. _____

20. Ⓟ = _____
    Ⓠ = _____

UNIT 16
Tolerances: Fractional and Angular Dimensions

TOLERANCES

As a part is planned, the designer must consider (1) its function either as a separate unit or as a part which must move in a fixed position in relation to other parts, (2) the operations required to produce the part, (3) the material to be used, (4) the quantity to be produced, and (5) the cost. Each of these factors influences the degree of accuracy to which a part is machined.

The dimensions given on a drawing are an indication of what the limits of accuracy are. These limits are called *tolerances.* If a part does not require a high degree of accuracy, the drawing may specify the tolerances to which the part is to be held in terms of *fractional dimensions.* More precisely machined parts require that the accuracy be given in terms of *decimal tolerances.*

SPECIFYING FRACTIONAL TOLERANCES

The note in figure 16-1, LIMITS ON FRACTIONAL DIMENSIONS ARE ± 1/64'', indicates that the dimension given in fractions on the drawing may be machined any size between a 64th of an inch larger to a 64th of an inch smaller than the specified size.

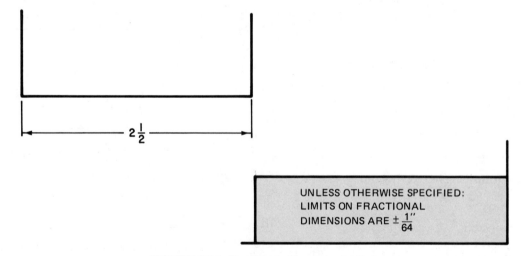

FIGURE 16-1 Fractional tolerances applied.

For example, on the 2 1/2'' dimension in figure 16-1:

1. The tolerance given on the drawing is ± 1/64''.

2. The largest size to which the part may be machined is 2 1/2'' + 1/64'' = 2 33/64''.

3. The smallest size to which the part may be machined is 2 1/2'' – 1/64'' = 2 31/64''.

The larger size is called the *upper limit;* the smaller size is called the *lower limit.*

ANGULAR DIMENSIONS

Angles are dimensioned in degrees or parts of a degree.

1. Each degree is one three hundred sixtieth of a circle (1/360).

2. The degree may be divided into smaller units called *minutes.* There are 60 minutes in each degree.

3. Each minute may be divided into smaller units called *seconds.* There are 60 seconds in each minute.

To simplify the dimensioning of angles, symbols are used to indicate degrees, minutes and seconds, figure 16-2. For example, twelve degrees, sixteen minutes and five seconds can also be written 12°16'5''.

	Symbol
Degrees	°
Minutes	'
Seconds	''

FIGURE 16-2 Symbols used for dimensioning angles.

SPECIFYING ANGULAR TOLERANCES

The tolerance on an angular dimension may be given in a note on the drawing as shown in figure 16-3. The tolerance may also be shown on the angular dimension itself, figure 16-4.

UPPER LIMIT: 45° + 30' = 45° 30'

LOWER LIMIT: 45° – 30' = 44° 30'

UNLESS OTHERWISE SPECIFIED:
LIMITS ON ANGULAR DIMENSIONS
ARE ± $\frac{1}{2}$°

FIGURE 16-3 Tolerance specified as a note.

TOLERANCE: ±10'

UPPER LIMIT: 60° + 10' = 60° 10'

LOWER LIMIT: 60° – 10' = 59° 50'

FIGURE 16-4 Tolerance specified on angular dimension.

60°

1 7/8

3/4

7/8

1/8 R

2 3/4

1/8 R

C

F

.250 REAM,
9/16 C'BORE, I DEEP

1/8 R

1 1/4

7/16

27/32

4 3/8

2 1/4

B

E

D

G

A

UNLESS OTHERWISE SPECIFIED:	ORDER NO. 29M 1602	PART NO. 29M 17	QTY. 2
LIMITS ON FRACTIONAL DIMENSIONS $\pm \frac{1}{64}$	MAT'L SAE 1090	HEAT TREATMENT SPEC. H 2	
LIMITS ON ANGULAR DIMENSIONS $\begin{array}{c}+\ 5'\\-\ 0'\end{array}$	**SECTOR PUNCH**		**BP-16**

SECTOR PUNCH (BP-16)

1. What letter is used to denote a

   (a)   Hidden edge line

   (b)   Dimension line

   (c)   Extension line

   (d)   Center line

2. Determine dimension Ⓖ

3. What is the diameter of the reamed pilot hole for the counterbore?

4. Give the diameter and depth of the counterbore.

5. What are the upper and lower limits of tolerance for fractional dimensions?

6. What limit of tolerance is specified for angular dimensions?

7. What is the largest size to which the 2 3/4″ diameter can be turned?

8. What is the lower limit to which the 2 3/4″ diameter can be machined?

9. Give the upper limit to which diameter Ⓑ may be machined.

10. Give the upper and lower limit on the diameter for shank Ⓕ.

11. If shank Ⓕ is turned to the upper limit length, how long will it be?

12. If shank Ⓕ is machined 2 3/16″ long, how much under the lower limit size will it be?

13. How much over the upper limit will the shank be if it is 2 5/16″ long?

14. What is the upper limit of accuracy for the 60° angle?

15. If the height of the 1 7/8″ diameter punch measures 1 9/32″, is it over, under or within the specified limits of accuracy?

ASSIGNMENT UNIT 16

Student's Name _____

1. (a) _____

   (b) _____

   (c) _____

   (d) _____

2. _____

3. _____

4. Diameter _____

   Depth _____

5. Upper _____

   Lower _____

6. _____

7. _____

8. _____

9. _____

10. Upper _____

    Lower _____

11. _____

12. _____

13. _____

14. _____

15. _____

UNIT 17
Tolerances: Unilateral and Bilateral Tolerances, and Decimal Dimensions

UNILATERAL AND BILATERAL TOLERANCES

The limits of accuracy to which a part is to be produced may fall within one of two classifications of tolerances. A dimension is said to have a *unilateral* (single) tolerance when the total tolerance is in one direction only, either (+) or (–). The examples of unilateral tolerances shown in figure 17-1 indicate that part (A) meets standards of accuracy when the basic dimension varies in one direction only and is between 3″ and 3 1/64″; part (B) may vary from 2.44″ to 2.43″; and angle (C) may vary between 60° and 59°45′.

FIGURE 17-1 Examples of unilateral tolerances.

Bilateral tolerances applied to dimensions mean that the dimensions may vary from a larger size (+) to a smaller size (–) than the basic dimension (nominal size). In other words, the basic dimension may vary in both directions. The basic 3″ dimension in figure 17-1A with a ± 1/64″ tolerance may vary between 3 1/64″ and 2 63/64″. The basic 2.44″ dimension in figure 17-1B with a bilateral tolerance of ± .01″ is acceptable within a range of 2.45″ and 2.43″. The 60° angle in figure 17-1C with a tolerance of ± 15′ may range between 60°15′ and 59°45′. These bilateral tolerances are shown in figure 17-2.

FIGURE 17-2 Examples of bilateral tolerances.

When the dimensions within the tolerance limits appear on a drawing, they are expressed as a range from the smaller to the larger dimension as indicated in figure 17-2.

DECIMAL DIMENSIONS

The decimal system of dimensioning is widely used in industry because of the ease with which computations can be made. In addition, the dimension can be measured with precision instruments to a high degree of accuracy.

68

Dimensions in the decimal system can be read quickly and accurately. Dimensions are read in thousandths, 1/1000″ = (.001″), in ten thousandths, 1/10,000″ = (.0001″), and in even finer divisions if necessary.

SPECIFYING DECIMAL TOLERANCES

Decimal tolerances can be applied in both the English and the metric systems of measurement. Tolerances on decimal dimensions which are expressed in terms of two, three, four, or more decimal places may be given on a drawing in several ways. One of the common methods of specifying a tolerance that applies on all dimensions is to use a note, figure 17-3.

FIGURE 17-3 Decimal tolerances given in a note.

UNLESS OTHERWISE SPECIFIED:
LIMITS ON DECIMAL
DIMENSIONS ARE ±.005″

1.750

For example, the 1.750″ dimension in the figure may be machined to a size ranging from

$$1.750″ + .005″ = 1.755″$$

TO

$$1.750″ - .005″ = 1.745″$$

The larger size (1.755″) is called the *upper limit*. The smaller size (1.745″) is called the *lower limit*.

A tolerance on a decimal dimension also may be included as part of the dimension, as shown in figure 17-4.

FIGURE 17-4 Decimal tolerance included with dimension.

1.750 ±.002

UPPER LIMIT   1.750″ + .002″ = 1.752″
LOWER LIMIT   1.750″ - .002″ = 1.748″

Bilateral tolerances are not always equal in both directions. It is common practice for a drawing to include either a (+) tolerance or a (−) tolerance that is greater than the other.

In cases where the plus and minus tolerances are not the same, such as plus .001″ and minus .002″, the dimension may be shown on the drawing as in figure 17-5.

FIGURE 17-5 Variation in limits.

1.750 $^{+.001}_{-.002}$

UPPER LIMIT   1.750″ + .001″ = 1.751″
LOWER LIMIT   1.750″ - .002″ = 1.748″

This same variation between the upper and lower limit can be given as in figure 17-6. The dimension above the line is the upper limit; the dimension below the line is the lower limit.

FIGURE 17-6 Upper and lower limit dimension.

1.751
1.748

UPPER LIMIT   1.751″
LOWER LIMIT   1.748″

ALL ROUNDED CORNERS HAVE .25 R

$\frac{31}{64}$ DRILL, .500 REAM
$\frac{7}{8}$ C BORE, .38 DEEP

UNLESS OTHERWISE SPECIFIED
TOLERANCES ARE:
FRACTIONAL DIMENSIONS ± $\frac{1}{64}$"

DECIMAL   TWO-PLACE: ±.01"
DIMENSIONS: THREE-PLACE: ±.001"
ANGULAR DIMENSIONS: ± 5'

QUANTITY: 15 REQUIRED
ORDER NO: 5 B 1072

DOVETAILED SLIDE   **BP-17**

MATERIAL:
GRAY CAST IRON

SCALE: $\frac{1''}{2} = 1''$

DOVETAILED SLIDE (BP-17)

1. What type of lines are Ⓑ, Ⓒ, Ⓗ, Ⓔ and Ⓙ?

2. What tolerance is allowed on
   (a) Fractional dimensions?
   (b) Decimal dimensions?
   (c) Angular dimensions?

3. What is the minimum overall height or thickness?

4. Give the upper limit dimension for the 60° angle.

5. Give the upper and lower limit dimensions for Ⓟ.

6. What is the maximum depth to which the counterbored hole can be bored?

7. What line in the top view represents surface Ⓡ of the side view?

8. What line in the front view represents surface Ⓛ?

9. What line in the side view represents surface Ⓐ of the top view?

10. What dimension indicates how far line Ⓙ is from the base of the slide?

11. What two lines in the top view indicate the opening of the dovetail?

12. How wide is the opening in the dovetail?

13. At what angle to the horizontal is the dovetail cut?

14. Give dimension Ⓨ.

15. To what depth into the piece is the dovetail cut?

16. What is the vertical distance from surface Ⓠ to surface Ⓣ?

17. What is the upper limit dimension between surfaces Ⓕ and Ⓖ?

ASSIGNMENT UNIT 17

Student's Name _____

1. _____ 11. _____

2. (a) _____ 12. _____
   (b)
   (c) _____ 13. _____

3. _____ 14. _____

4. _____ 15. _____

5. _____ 16. _____

6. _____ 17. _____

7. _____ 18. _____

8. _____ 19. _____

9. _____ 20. _____

10. _____ 21. _____

22. (a) _____
    (b) _____
    (c) _____

18. What is the full depth of the tee slot?

19. Compute dimensions Ⓥ and Ⓧ.

20. What is the horizontal distance from line Ⓝ to line Ⓢ?

21. What classification of tolerances applies to all dimensions?

22. Change the fractional, decimal, and angular tolerances so that only the (–) tolerances apply. Then, determine the upper and lower limit dimensions for:
    (a) The angular dimension
    (b) The distance between surfaces Ⓕ and Ⓖ.
    (c) Distance Ⓟ.

UNIT 18
Representing and Dimensioning
External Screw Threads

Screw threads are used widely (1) to fasten two or more parts securely in position, (2) to transmit power such as a feed screw on a machine, and (3) to move a scale on an instrument used for precision measurements.

PROFILE OF UNIFIED AND THE
AMERICAN NATIONAL THREADS

The shape or profile of the thread is referred to as the *thread form*. One of the most common thread forms resembles a V. Threads with an included angle of 60° originally were called Sharp V. Later, these threads were called U.S. Standard, American National, and Unified threads. Changes and improvements in the thread forms have been made over the years through standards which have been established by professional organizations. These include the National Screw Thread Commission, the American Standards Association, the U.S. Standards Institute and (currently) the American National Standards Institute (ANSI). The newer Unified (UN) thread form and sizes are designed to correct production problems related to tolerances and classes of fit. The *Unified Thread System* was agreed upon by Canada, the United States, and the United Kingdom.

The most commonly used thread forms are the *Unified* and the *American National*. UN threads are mechanically interchangeable with the former American National threads of the same size and pitch. The symbol UN is found on drawings to designate the United form, and N designates the American National form. The only difference between these two forms is in the shape of the top (crest) and the bottom (root) of the thread. Both the crest and root of the American National thread form are flat. In contrast, the crest and root of the Unified thread form may be either flat or rounded. The root is always rounded on UNR external threads. The characteristics of the basic profile for American National and Unified Screw Threads are shown in figure 18-1.

FIGURE 18-1 Characteristics of the American National Standard profile for Unified Screw Threads (N, UN, and UNR)

AMERICAN NATIONAL STANDARD AND UNIFIED SCREW THREADS

There are five common thread forms which use the 60° included angle. These forms are represented in figure 18-2 as (A) the Sharp V, (B) Unified, UN, (C) American Standard, N, (D) 60° Stub, and (E) ANSI Straight, NPS, and Taper, NPT, Pipe Threads. While flat crests and roots are shown for the pipe threads, some rounding off occurs in manufacturing.

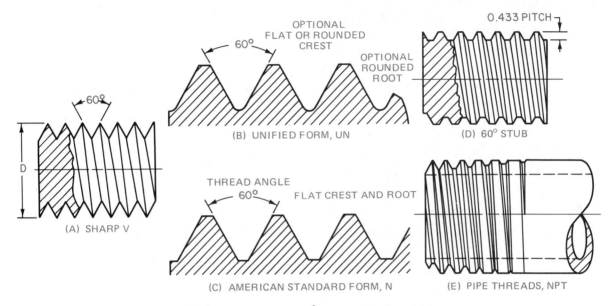

FIGURE 18-2 Common 60° angle screw thread forms.

OTHER COMMON THREAD FORMS

A square form of thread, or a modified form, is used when great power or force is needed. The square form shown in figure 18-3A is called a *Square thread.* The more popular modified thread form has an included angle of 29° and is known as the *Acme thread,* figure 18-3B. The *Buttress thread,* figure 18-3C, is used where power is to be transmitted in one direction. The load-resisting flank may be inclined from 1° to 5°.

FIGURE 18-3 Square and modified thread forms.

THREAD SERIES

Each thread form has a standard number of threads per inch for a given diameter. The thread series designates the fineness or coarseness of the threads. For example, there are five basic series of 60° thread form screw threads, as follows.

(1) The *coarse thread series* is designated as UNC for Unified Coarse; UNR for Unified Coarse having a specified rounded root but optional rounded crest; and NC for the American National Coarse Series threads.

(2) The *fine thread series* is similarly identified by the symbols UNF, UNRF for external threads, and NF.

(3) *Extra fine series threads* are designated by the symbols UNEF, UNREF (for external threads only), and NEF.

(4) *Nonstandard* or *special threads* are designated by the symbol UNS for Unified Special and NS for American National Standard Special Series threads.

(5) A *constant pitch series* provides a fixed number of threads (pitch) for a range of diameters. The 8, 12, and 16 constant pitch series are preferred. Threads in each series are designated as UN, UNR (for external threads), and N. These letters (depending on the thread form) are preceded by the number of threads in the constant pitch series.

Figure 18-4 provides a summary of the five thread series and the designations as they appear on drawings.

	Designations on Drawings		
Thread Series	**Unified System**		**American Standard National System**
Coarse	∠1 UNC	∠2 UNRC	NC
Fine	UNF	UNRF	NF
Extra Fine	UNEF	UNREF	NEF
Special	UNS	UNRS	NS
Constant Pitch	∠3 * UN	* UNR	* N

∠1 Optional rounded crest and/or root
∠2 Specified root radius; optional rounded crest
∠3 *Pitch series number precedes the designation

FIGURE 18-4 Summary of thread series designations.

REPRESENTING EXTERNAL SCREW THREADS

Screw threads are further classified into two basic types: (1) *external threads* which are produced on the outside of a part, and (2) *internal threads* which are cut on the inside of the part. Letter symbols are sometimes used to designate the type: A for external threads and B for internal threads.

Internal and external threads may be represented on mechanical drawings by one of several methods: (1) a *pictorial* representation which shows the threads as they appear to the eye, (2) a *schematic* representation, or (3) a *simplified* representation. Figure 18-5 shows how each type of external thread can be represented.

FIGURE 18-5 Representation of external screw threads (ANSI).

DIMENSIONING EXTERNAL SCREW THREADS

The representation of each thread is accompanied by a series of dimensions, letters, and numbers which, when combined, give full specifications for cutting and measuring the threads. The standard practices recommended by the American National Standards Institute (ANSI) for specifying and dimensioning external screw threads are shown in figure 18-6.

While the specifications as noted in figure 18-6 give all the information needed to describe a screw thread, the seven items are not always used in the notation on a drawing. For example, the thread class ⑤ may be covered by a general note which applies to all threaded parts. Such a note may appear elsewhere on the drawing. In other cases, the length of thread ⑦ sometimes appears as a dimension instead of appearing in the thread notation.

DIMENSIONING CODE

① OUTSIDE (NOMINAL) DIAMETER OF THREADS

② NUMBER OF THREADS PER INCH

③ THREAD FORM (NATIONAL)

④ THREAD SERIES (COARSE)

⑤ THREAD CLASS

⑥ EXTERNAL THREAD

⑦ LENGTH OF THREAD

FIGURE 18-6 Dimensioning external screw threads.

PIPE THREAD REPRESENTATION

In addition to the screw threads just covered, there are three other common types of ANSI threads that are used on pipes. One type is the *Straight pipe thread*. This type of thread is noted on a drawing by the diameter of the thread followed by NPS. Thus, the notation, 1″ - NPS, indicates that (a) the part contains the standard number of threads in the pipe series for a 1″ diameter, and (b) the threads are straight.

In the second type of pipe thread, the threads are cut along a standard taper. The tapered pipe threads are used where a tight, leakproof seal is needed between the parts that are joined. Such threads are called *National Pipe Taper* threads, designated by NPT. They are specified on pictorial, schematic, or simplified drawings by the diameter and the type. The notation, 1″ - NPT, means (a) that there are the standard number of threads on the part for the 1″ diameter in the pipe thread series, and (b) both the internal and external threads are cut on the standard pipe taper.

A third type of pipe thread is used on drawings of parts for extremely high pressure applications. *Dryseal pressure-tight threads* are identified by the addition of the letter P after the N. For example, *NPTF* designates threads in the *Dryseal taper series; NPSF,* threads in the *Dryseal straight pipe thread series.*

While a number of different forms and thread series have been described, only the commonly used American National Standard (ANSI), Unified System, and International Standards Organization (SI) Metric Thread System are applied in this basic text.

$\frac{3}{32}$ DRILL AND 60° COUNTERSINK
TO .25 DIAMETER – BOTH ENDS

1–14 UNF

NECK TO THREAD DEPTH

$1.125 ^{+.000}_{-.000}$

$\frac{1}{8}$ R

.750

1.562

.500

2.25 ±.01

UNDERCUT $\frac{1}{16}$

1.250 $^{+.000}_{-.001}$

Ⓒ

$1\frac{1}{4}$ – 12 UNRF

1.125

$\frac{5}{8}$

$\frac{1.062}{1.061}$

Ⓐ

1.000

$\frac{.750}{.749}$

Ⓑ

Ⓓ

1.500

$\frac{3}{4}$ – 10 UNC
$\left(^{+.000}_{-.002}\right)$

$\frac{7}{8}$

30° CHAMFER
TO THREAD DEPTH
FOR ALL THREADS

GRIND ALL DIAMETERS
AFTER HARDENING

UNLESS OTHERWISE SPECIFIED
LIMITS ON DIMENSIONS ARE:

FRACTIONAL DIMENSIONS $+\frac{1''}{64}$, –0''

DECIMAL DIMENSIONS ± .002''

ANGULAR DIMENSIONS + 1°, –0°

MAT'L. SAE 2335

H. T. HARDEN AND TEMPER

QUANTITY 42

ORDER NO. L 224 – 1

SPINDLE SHAFT   **BP-18**

SPINDLE SHAFT (BP-18)

1. What material is used for the part?

2. Give the overall length of the shaft.

3. What system of representation is used for the threaded portions?

4. At how many places are threads cut?

5. Start at the bottom of the part and give all the thread diameters.

6. Name the three thread series that the letters UNC, UNRF, and NC specify.

7. How many threads per inch are to be cut on the 3/4", 1 1/4" and 1" diameters?

8. Give dimensions Ⓑ, Ⓒ and Ⓓ .

9. Give the upper and lower limit dimensions for the 3/4" threaded portion.

10. What is the length of the 3/4" threads? The 1 1/4" threads?

11. Give the upper and lower limit dimensions of the 1 1/16" diameter portion.

12. What angle are the chamfers at the starting end of each thread cut?

13. What tolerance is specified for angular dimensions?

14. What is the upper and lower limit of size on the 1 1/8" diameter?

15. How long is that part of the Shaft which has the 1 1/4" - 12 thread?

16. Give the angle of the countersink.

17. What is the largest diameter that the 2.25" diameter can be machined to?

18. Give the diameter to which the center holes are countersunk.

19. How deep is the undercut on the 1 1/4 diameter?

20. Name the final machining operation for all diameters after the part is heat treated.

21. Classify the tolerances for the four ground diameters.

ASSIGNMENT UNIT 18

Student's Name _____

1. _____   9. Upper _____
2. _____   Lower _____
3. _____   10. 3/4" _____
   _____   1 1/4" _____
4. _____   11. Upper _____
5. _____   Lower _____
   _____   12. _____
   _____   13. _____
6. UNC _____   14. Upper _____
   _____   Lower _____
   UNRF _____   15. _____
   NC _____   16. _____
   _____ 17. _____
7. 3/4" _____   18. _____
   1 1/4" _____
   1" _____   19. _____
8. Ⓑ _____   20. _____
   Ⓒ _____
   Ⓓ _____   21. _____

UNIT 19
Representing and Specifying Internal and Left-hand Threads

REPRESENTATION OF INTERNAL THREADS

One of the most practical and widely used methods of producing internal threads is to cut them with a round, formed thread-cutting tool called a *tap*. The tap is inserted in a hole which is the same diameter as the root diameter of the thread to be produced. As the tap with its multiple cutting edges is turned, it advances into the hole to cut a thread. The thread is a specified size and shape to correspond with a mating threaded part. This process is called *tapping* and the hole which is threaded is called a *tapped hole*. When the hole is a drilled hole, the drill is called a *tap drill.*

A part may be threaded internally either throughout its entire length or to a specified depth either by tapping or another machining process. Some threads are bottomed at the same depth as the tap drill. In other cases, the tap drill hole may be deeper. Pictorial, schematic and simplified forms of representing such threads are used in figure 19-1. Section (A) illustrates through threads; Section (B) illustrates threads in a blind hole.

FIGURE 19-1 Representation of internal threaded holes.

DIMENSIONING A THREADED HOLE

The same system of dimensioning that is used to specify external threads applies to internal threads. However, in the case of internal threads, the thread length is specified as the *depth of thread*. The recommended dimensioning of a tapped hole is illustrated in figure 19-2.

①	2″ – OUTSIDE THREAD DIAMETER
②	8 – 8 PITCH SERIES
③	N – THREAD FORM
④	2 – CLASS 2 THREAD FIT
⑤	B – INTERNAL THREAD
⑥	LH – LEFT-HAND SCREW THREAD
⑦	$3\frac{1}{8}$ – DEPTH OF THREADS

FIGURE 19-2 Dimensioning a tapped hole.

THREAD CLASS: UNIFIED AND AMERICAN NATIONAL FORM THREADS

The accuracy between an external and an internal thread is specified in trade handbooks and manufacturer's tables as a *thread class*. In the old classes of thread fits, there were equal allowances (tolerances) on the pitch diameters of both the external and internal threads. Today, the new standards require a larger pitch diameter tolerance on the internal thread.

There are three regular classes of fits for external and internal threads. Each of the three classes is distinguished from one another by the amount of clearance between the sides, crests, and roots of the internal and external threads.

The variation is from the loose *Class 1* fit to *Class 3* which requires a high degree of accuracy. Fits that are Class 1A (where A designates external threads) and Class 1B (B for internal threads) provide the greatest allowance and tolerances between the fitted parts. The Class 2A and Class 2B fit is the most widely used as it provides an acceptable minimum of clearance between the mating parts. The Class 3A and Class 3B fit is used for great accuracy. When the class of fit is given, it is usually included on the drawing as part of the screw thread notation, figure 19-2. The current thread classes 1, 2, and 3 replace the former American National Standard 1, 2, and 3 classes of fit.

SPECIFYING LEFT-HAND THREADS

Screw threads are cut either right-hand or left-hand depending on the application, figure 19-3. As the terms imply, a right-hand thread is advanced by turning clockwise ⟲ or to the right; the left-hand thread is advanced by turning counterclockwise ⟳, or to the left.

FIGURE 19-3 Examples of right- and left-hand threads.

No notation is made on a drawing for right-hand threads as this direction of thread is assumed, unless otherwise noted. However, the letters LH are included on the drawing specifications to indicate when a left-hand thread is required as shown in figure 19-2 by the LH notation at ⑥.

$\frac{55}{64}$ DRILL, .875 REAM

$\frac{5}{16}$ DRILL, .56 DEEP

TAP $\frac{5}{16}$ – 18 NC – 1

.82

.12

1.625 $^{+.002}_{-.000}$

C

E

3.00

1.50

.82

.12

1.25

2.200
2.198

B

1.250
1.248

$\frac{3}{4}$ – 16 UNRF – 2LH

.62

1.38

NO. 10 – 32 UNF, .44 DEEP
4 HOLES

.750

$\frac{31}{64}$ DRILL, .500 REAM
2 HOLES

.12 R

A

.88

.75

45°

1.500
1.499

.31

.50

.750
.749

1.25

D

1 $\frac{1}{2}$ – 16 N – 3 LH

ALL FILLETS AND ROUNDS .12 R

ORDER NO.	QTY.	MAT'L.	H.T.
9 – H 3	5	CAST STEEL	NORMALIZE AFTER ROUGH MACHINING

UNLESS OTHERWISE SPECIFIED
LIMITS ON DIMENSIONS ARE:

FRACTIONAL DIMENSIONS ± $\frac{1}{64}$"

DECIMAL DIMENSIONS: ± .01" (TWO-PLACE);
± .001" (THREE-PLACE)

ANGULAR DIMENSIONS ± 15'

ROCKER BASE   **BP-19**

DR. Judith Forward

CK. C. G. Whitehurst

PART NO. Z – 721

DATE 9-14-82

ROCKER BASE (BP-19)

1. What material is specified?

2. What tolerances are allowed on:
   (a) Decimal dimensions?
   (b) Fractional dimensions?
   (c) Angular dimensions?

3. What heat-treating process is required after rough machining?

4. Determine overall dimensions (A) (B) (C).

5. What is the radius of all fillets and rounds?

6. Compute dimensions (D) and (E).

7. How many holes are to be reamed .500"?

8. How many external threads are to be cut?

9. What thread series is used for the internal and external threads?

10. What does 3/4 - 16UNRF - 2LH mean?

11. What are the upper and lower limits of the .750" diameter portion?

12. What does 1 1/2 - 16N - 3LH mean?

13. What does 5/16 - 18NC - 1 mean?

14. What is the lower limit diameter of the unthreaded 1 1/2" diameter portion?

15. Determine the length of the 1 1/2" - 16 threaded portion.

16. What diameter drill is used for the .875" reamed hole?

17. How many holes are to be threaded 10 - 32UNF?

18. How deep are the holes to be threaded?

19. Give the angle to the horizontal at which the 10 - 32UNF holes are to be drilled.

20. Give the diameter of the circle on which the 10 - 32UNF tapped holes are located.

ASSIGNMENT UNIT 19

Student's Name _____

1. _____   11. Upper _____
                            Lower _____
2. (a) _____
   (b) _____   12. 1 1/2 _____
   (c) _____   16 _____
                         N _____
3. _____
                         3 _____
4. (A)_____   LH _____
   (B)_____
   (C)_____   13. 5/16 _____
                        18 _____
5. _____   NC _____
6. (D)_____   1 _____
   (E)_____
                    14. _____
7. _____
                    15. _____
8. _____
                    16. _____
9. _____
                    17. _____
10. 3/4 _____
    16 _____   18. _____
    UNRF _____
    2 _____   19. _____
    LH _____   20. _____

21. _____

22. (A)_____ (D) _____
    (B)_____ (E) _____
    (C)_____ 45°⤡_____

21. Identify the system of dimensioning used on the drawing of the Rocker Base.

22. Change the fractional, decimal and angular tolerances so they apply in one direction (+) only. Then, determine the upper and lower limit dimensions for (A) (B) (C) (D) (E) and the 45° angle.

UNIT 20
Dimensioning Tapers and Machined Surfaces

Blueprints of parts that uniformly change in size along their length show that the part is *tapered.* On a round piece of work, the taper is the difference between the diameter at one point and a diameter farther along the length. The taper is usually specified on drawings where dimensions are given in inches by a note which gives the TAPER PER FOOT or the TAPER PER INCH. For example, a taper of one-half inch per foot is expressed:

$$\frac{1}{2} \text{ TAPER PER FOOT}$$

The drawing may give the large and small diameters at the beginning and end of the taper or one diameter and the length of the taper, figure 20-1.

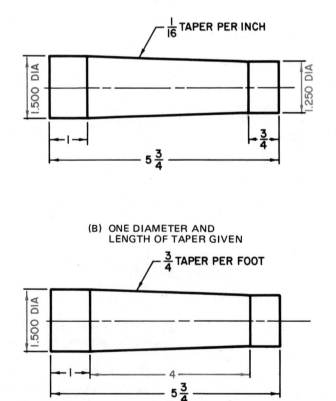

(A) TWO DIAMETERS GIVEN

(B) ONE DIAMETER AND LENGTH OF TAPER GIVEN

FIGURE 20-1 Notes used in dimensioning tapers.

FINISHED SURFACES

Surfaces that are machined to a smooth finish and to accurate dimensions are called *finished surfaces.* These surfaces were identified by the former 60° ∨ ANSI symbol, the √ symbol which was accepted by the Canadian Standards Association (CSA) and the ANSI, and the CSA symbol *f*. Presently, the symbol ▽ is used to indicate that material is to be removed by machining the designated surface.

The bottom of the finish symbol touches the surface to be machined. Figure 20-2 shows a steel forging. This same forging as it looks when machined is illustrated in figure 20-3. The working drawing, figure 20-4, shows how the finish marks are placed. Note that these marks may appear on the hidden edge lines as well as on the object lines, with the exception of the small drilled holes. On large holes that require accurate machining, the ▽ is used once again. Extension lines and/or leaders are also used with the finish symbol to identify surfaces which are to be machined.

FIGURE 20-2 Rough forging.

FIGURE 20-3 Machined forging.

FIGURE 20-4 Working drawing of forging showing application of finish marks.

FINISH ALL OVER

When a casting, forging, or welded part is to be finished all over, the drawing is simplified by omitting the finish symbols and adding the note: FINISH ALL OVER. The abbreviation FAO may be used in place of the phrase, finish all over.

ANSI SYMBOLS FOR SURFACE FINISH

Four basic ANSI symbols for denoting the process by which a surface may be finished are shown in figure 20-5, column (a). A brief description of the process which each symbol identifies on a drawing is given in column (b).

ANSI Surface Finish Symbol (a)	General Processes Identified on Drawings (b)
$\sqrt{}$	The surface may be produced by any manufacturing process
$\sqrt{}$	The surface is to be produced by a nonmachining process such as casting, forging, welding, rolling, powder metallurgy, etc. There is to be no subsequent machining of the surface.
$\sqrt{}$	Material removal from the surface by machining is required.
.05″ $\sqrt{}$ or 1.5mm $\sqrt{}$	The addition of a decimal or metric value indicates the amount of stock to be removed by machining.

FIGURE 20-5 Basic ANSI surface finish symbols.

REPRESENTATION AND MEASUREMENT OF SURFACE TEXTURE

The quality of a machined surface, particularly where a high degree of precision is required, is identified on a drawing by the symbol $\sqrt{}$. The ANSI surface texture system requires a series of measurement values to be used with the symbol. The values specify the degree of accuracy to which surface irregularities (*roughness*) are to be measured. The pattern of marks produced by machining (*lay*), the spacing of the irregularities (*waviness*), and scratches and other surface imperfections (*flaws*) are represented by still other symbols and measurement values. These are also added in designated positions to the basic finish symbol.

The term *surface texture,* as applied to drawings, denotes the allowable variations in roughness, waviness, lay, and/or flaws. Surface texture symbols and values specify precise standards against which finished parts may be inspected and measured.

1.125

.750 TAPER PER FT.

.12

.94

C

B

H

1⅞   2¾

.500 DIA

G

.375

1 21/64

OFFSET ARM

.62

45°

¼ DRILL, 15/32 C'BORE
5/16 DEEP

39/64

¾ R

¾ R

¾ R

A

.25

5/8

1

3/16 DRILL
90° CSK. TO ½ DIA
4 HOLES

1½

PAD

33/64

E

F

1 R

7/8 R

.12 R

D

BODY

UNLESS OTHERWISE SPECIFIED TOLERANCES ON DIMENSIONS ARE:	MAT'L.	PART NO.	QTY. 650
DECIMAL DIMENSIONS: ± .01" (TWO-PLACES) ± .002" (THREE-PLACES)	SAE 1040	FORGING DF – 10 – 625	HEAT TREAT. NORMALIZE
FRACTIONAL DIMENSIONS ± 1"/64 ANGULAR DIMENSIONS ± 15'		**OFFSET CARRIER ARM**	**BP-20**

OFFSET CARRIER ARM (BP-20)

1. Name the three views.

2. What is the part number?

3. How many surfaces are to be machined?

4. Give the angle at which the arm is offset from the body.

5. What is the center-to-center distance of the offset arm?

6. Find the length of the elongated slot.

7. What is the overall length of the pad?

8. How is the counterbored hole specified?

9. What diameter drill is used for the countersunk holes?

10. What is the angle of the countersunk holes?

11. Compute the diameter at the large end of the tapered hole (G).

12. What tolerance is allowed on:
    (a) Decimal dimensions?
    (b) Angular dimensions?
    (c) Fractional dimensions?

13. What is the maximum overall length (A)?

14. Determine the minimum overall width (B).

15. Determine the maximum overall thickness (C).

16. What are dimensions (D) and (H)?

17. If the elongated slot is machined 1/2", would it be over, under, or within the specified limits?

18. What is the distance (E)?

19. Give center-to-center distance (F).

20. What heat treatment is required?

21. What classification of tolerances applies to all dimensions?

ASSIGNMENT UNIT 20

Student's Name _____

1. _____   12. (a) _____

                        (b) _____

                        (c) _____

2. _____   13. (A) = _____

3. _____   14. (B) = _____

4. _____   15. (C) = _____

5. _____   16. (D) = _____

6. _____   (H) = _____

7. _____   17. _____

8. _____   18. (E) = _____

                   19. (F) = _____

9. _____   20. _____

10. _____   21. _____

11. (G) = _____

22. (A) = Min. _____ Max. _____
    (B) = Min. _____ Max. _____
    (C) = Min. _____ Max. _____
    (D) = Min. _____ Max. _____
    (E) = Min. _____ Max. _____
    (F) = Min. _____ Max. _____
    (H) = Min. _____ Max. _____

22. Use the appropriate fractional or decimal tolerance so it applies in one direction (+) only. Then, determine the minimum and maximum limit dimensions for (A) (B) (C) (D) (E) (F) and (H).

UNIT 21
Dimensioning with Shop Notes

The draftsperson often resorts to the use of notes on a drawing to convey to the mechanic all the information needed to make a part. Notes such as those used for drilling, reaming, counterboring, or countersinking holes are added to ordinary dimensions.

A note may consist of a very brief statement at the end of a leader, or it may be a complete sentence which gives an adequate picture of machining processes and all necessary dimensions. A note is found on a drawing near the part to which it refers. This unit includes the types of machining notes that are found on drawings of knurled surfaces, chamfers, grooves, and keyways. A sample of a typical change note is also given.

DIMENSIONING KNURLED SURFACES

The term *knurl* refers to a raised diamond-shaped surface or straight line impression in the surface of a part, figure 21-1A. The dimensions and notes which furnish sufficient information for the technician to produce the knurled part are also given, figure 21-1B and figure 21-1C. The pitch of the knurl, which gives the number of teeth per linear inch, is the size. The standard pitches are: coarse (14P), medium (21P), and fine (33P).

Newer ANSI standards for knurls and knurling use the letter P to precede the pitch. Where parts are to be press fit, the minimum acceptable (toleranced) diameter is indicated on the drawing. A knurled surface is represented by diamond or straight line pattern symbols or by using the designation of the knurl as straight or diamond. Common methods of representing knurls are shown in figure 21-1B and figure 21-1C. A partial pattern is shown when the symbol clarifies a drawing.

FIGURE 21-1 Patterns, dimensions, and notes for knurled surfaces.

The pitch of a knurl is usually stated on drawings in terms of the number of teeth as shown in figure 21-1B or fractional part per millimeter, figure 21-1C. The pitch is followed by the type of knurl.

DIMENSIONING CHAMFERS AND GROOVES

When a surface must be cut away at a slight bevel, or have a groove cut into it, the drawing or blueprint gives complete machining information in the form illustrated in figure 21-2 and figure 21-3. The angle is dimensioned as an included angle when it relates to a design requirement.

FIGURE 21-2 Dimensioning chamfers and bevels.

FIGURE 21-3 Dimensioning three common types of grooves.

The process of producing a groove is often referred to as *undercutting* or *necking*. The simplified ANSI method of dimensioning a round or square (plain) undercut is to give the width followed by the diameter. For example, a note such as .12 x Ø 1.25 indicates a groove width of .12" and a 1.25" diameter at the bottom of the undercut. If a round groove is required, the radius is one-half the width dimension.

KEYS, KEYWAYS, AND KEYSEATS

Parts that are assembled and disassembled in a particular position to transmit motion from one member to another are *keyed* together. This means a square or flat piece of steel is partly seated in a shaft or groove called a *keyseat*. The key extends an equal distance into a corresponding groove *(keyway)* in a second member *(hub)*. Another general type of key is semicircular in shape. It fits into a round machined groove in the shaft. This type is known as a *Woodruff key.*

Dimensions and specifications as they appear on drawings for square and flat keys, keyseats, and keyways are shown in figure 21-4. Drawings often include a height dimension which gives the measurement from the circumference to the depth to which the keyway and keyseat are machined.

FIGURE 21-4 Key specifications and dimensioning keyseats and keyways.

CHANGE NOTES

The specifications and dimensions of parts are frequently changed on working drawings. An accurate record is usually made on the tracing and blueprint to indicate the nature of the change, the date, and who made the changes.

Changes that are minor in nature may be made without altering the original lines on the drawing. One of the easiest ways of making such changes is shown in figure 21-5.

FIGURE 21-5 Application of change notes.

Ø 2.75

2 – 16 N – 2

Ⓑ

30° CHAMFER TO
THREAD DEPTH

.625 TAPER PER FOOT

2.50

1.25

CHAMFER .06 x 45°

UNDERCUT .12R x .12 WIDE
TO THREAD DEPTH

NO. 7 DRILL, .32 DEEP
4 HOLES

.62

NOTE: 82° C'S'K. TO
.26 DIA AFTER KNURLING

14 P
DIAMOND
KNURL

Ⓐ

.50

REAM .875

.50

NECK .12 x .12 R
TO .12 DEEP

3.500±.001

Ⓘ

1.500 TAPER PER FOOT

◁G

Ⓒ

Ⓓ

NECK .12 x .12 R TO .01 DEEP

.250 x .125 KEYWAY

1.12

Ⓐ

.75

.12R

1.250 DIA
1.249 DIA

⅝ – 11 UNC – 1 LH – .88 DEEP

NO.	CHANGE	DATE	BY	
②	WAS 1⅛	5-3-82	T.P.O.	
①	.125 ADDED	5-3-82	T.P.O.	

UNLESS OTHERWISE SPECIFIED
TOLERANCES ON DIMENSIONS ARE:
DECIMAL DIMENSIONS ± .01" (TWO-PLACE):
± .002" (THREE-PLACE)
FRACTIONAL DIMENSIONS ± $\frac{1''}{64}$
ANGULAR DIMENSIONS ± 1°

CUTTER ADAPTER   BP-21

CUTTER ADAPTER (BP-21)

1. What tolerances are given for:
   (a) Angular dimensions?
   (b) Decimal dimensions?
   (c) Fractional dimensions?

2. Give the exact specifications for the taper on the body.

3. What is the taper per inch for the body?

4. What is the diameter at the small end of the outside taper?

5. Determine dimension (A).

6. Give the taper for the hole.

7. What is the taper per inch for the hole?

8. Give the diameter at the small end of the inside taper.

9. Compute dimension (B).

10. Give the specifications for the outside threads on the nose portion.

11. What does the LH in the thread note for the shank end specify?

12. What is the thread class for the 2″ threaded nose?

13. What operation is performed on the 2.75″ diameter.

14. How is this operation specified?

15. Determine the maximum diameter for the reamed hole.

16. Give the note for dilling the four holes in the largest diameter.

17. Specify the chamfer for the knurled portion.

18. Specify the angle of chamfer for the 2″ threaded nose.

19. How deep is the chamfer for the 2″ threads?

20. Give the original length of the tapered body before any change was made.

ASSIGNMENT UNIT 21

Student's Name _____

1. (a) Angular _____
   (b) Decimal _____
   (c) Fractional _____

2. _____   15. _____

3. _____   16. _____

4. _____   _____

5. _____   17. _____

6. _____   18. _____

7. _____   19. Deep _____

8. _____   Angle _____

9. _____   20. _____

10. _____   21. _____

11. _____   22. _____

12. _____   23. _____

13. _____   24. Width _____

14. _____   Depth _____

_____   25. _____

26. (A) _____

    (B) _____

21. Compute maximum length of dimension (C).

22. What machining operation is indicated by the finish symbol $\overset{G}{\forall}$?

23. Determine dimension (D).

24. Give the dimensions of the keyseat.

25. Determine the overall length of the adapter.

26. Change the decimal tolerance to apply unidirectionally as (+.003″). Then, compute the minimum and maximum dimensions for (A) and (B).

SECTION 4
The SI Metric System

UNIT 22
Metric System Dimensioning
Diametral Dimensions, and ISO Symbols

The International System of Units (SI) was established by agreement among many nations to provide a logical interconnected framework for all measurements used in industry, science, and commerce. SI is a modernized accepted adaptation of many European metric systems. SI metrics relates to six basic units of measurement: (1) length, (2) time, (3) mass, (4) temperature, (5) electric current, and (6) luminous intensity. Multiples and submultiples of these basic units are expressed in decimals. All additional SI units are derived from the six basic units.

THE SI METRIC SYSTEM OF MEASUREMENT

The SI metric system of measurement is gaining wide acceptance by industry, throughout America and the world. The system has been used by the chemical and photographic industries, the medical professions, and in other areas for many years. As the use of the system spreads, dimensions and other features on drawings will gradually change to units of measurement expressed as SI metric units or in combination with customary inch standard units.

Prefixes are used in the metric system to show how the dimension relates to a basic unit. For example, *deci-* means one-tenth of the basic unit of measure. Thus, 1 decimeter = 0.1 of a meter; 1 deciliter = 0.1 liter; and 1 decigram = 0.1 gram. Similarly, *centi-* = one hundredth; and *milli-* = one thousandth.

Dimensions larger than the basic unit of measure are expressed with the following prefixes: *deka-* = ten times greater; *hecto-* = one hundred times; and *kilo-* = 1000 times greater. The most common unit of metric measure used on drawings is the millimeter (mm). The next frequently used unit is the meter (m). Surface finishes are given in *micrometers* (millionths of a meter), using the symbol *μm*.

CONVERTING METRIC AND CUSTOMARY
INCH SYSTEM DIMENSIONS

Trade and engineering handbooks contain conversion charts which simplify the process of determining in one system the equivalent value of a dimension given in the other system. Sections of a conversion table are illustrated in figure 22-1 to show customary inch units and equivalent metric units. Figure 22-1 also includes a drill series ranging from #80 to #1, letter size drills from (A) to (Z), and fractional dimensions from .001″ to 1.000″. Note that millimeter equivalents, correct to four decimal places, are shown for both the drill sizes and the decimals which represent fractional parts of an inch.

The millimeter range of the table is from 0.0254 mm to 25.4000 mm, corresponding to 0.001″ to 1.000″, respectively. The millimeter equivalent of a measurement in the inch system may be found by multiplying the decimal value by 25.40.

94

Drill No. or Letter	Inch Standard	SI Metric Standard
	Fractional or Decimal Size	mm Size
	.001	0.0254
	.002	0.0508
	.003	0.0762
	.004	0.1016
	.005	0.1270
	.006	0.1524
	.007	0.1778
	.008	0.2032
	.009	0.2286
	.010	0.2540
	.011	0.2794
	.012	0.3048
80	.0135 .013	0.3302
79	.0145 .014	0.3556
	.015	0.3810
	1/64 .0156	0.3969
78	.016	0.4064
	.017	0.4318

Drill No. or Letter	Inch Standard	SI Metric Standard
	Fractional or Decimal Size	mm Size
	.401	10.1854
	.402	10.2108
	.403	10.2362
Y	.404	10.2616
	.405	10.2870
	.406	10.3124
13/32	.4062	10.3187
	.407	10.3378
	.408	10.3632
	.409	10.3886
	.410	10.4140
	.411	10.4394
	.412	10.4648
Z	.413	10.4902
	.414	10.5156
	.415	10.5410
	.416	10.5664

Drill No. or Letter	Inch Standard	SI Metric Standard
A	.234	5.9436
15/16	.2344	5.9531
	.235	5.9690
	.236	5.9944

Drill No. or Letter	Inch Standard	SI Metric Standard
	.996	25.2984
	.997	25.3238
	.998	25.3492
	.999	25.3746
	1.000	25.4000

FIGURE 22-1 Portions of a conversion table showing drill sizes and millimeter equivalents to inch dimensions.

ROUNDING OFF LINEAR DIMENSIONS

Many dimensions that are converted from inches to millimeters or from millimeters to inches are rounded off. With 1 mm = 0.03937", conversions can be made to very high limits of accuracy. This is done by calculating the decimal value to a greater number of digits than is required within the range of tolerances.

Production costs are related directly to the degree of accuracy required to produce a part. Therefore, the limits of dimensions should be rounded off to the least number of digits in the decimal dimension which will provide the greatest tolerance and ensure interchangeability.

A decimal dimension may be rounded off by increasing the last required digit by (1) if the digit which follows on the right is (5) or greater or by leaving the last digit unchanged if the digit to the right is less than (5).

For example, 1.5875"   rounded off to three decimal places = 1.588"
1.5874 mm   rounded off to three decimal places = 1.587 mm
1.646 mm   rounded off to two decimal places   = 1.65 mm
1.644"   rounded off to two decimal places   = 1.64"

When dimensions on a drawing need to be divided, the part designer may round off the dimensions to the nearest even decimal. This practice permits parts to be economically machined to tolerances within a required number of decimal places.

DIAMETRAL DIMENSIONS

The symbol Ø on a drawing indicates that the part has cylindrical sections. For example, the part shown in figure 22-2 may be fully described in one view by using the diametral dimension symbol Ø for the cylindrical surfaces. The one-view drawing gives the dimensions of each of the three cylindrical surfaces and the bore diameter. The Ø symbol may precede or follow the dimensional value.

FIGURE 22-2 Use of diametral dimension symbol.

PROJECTION SYMBOLS

The International Organization for Standardization (ISO) recommends the use of a projection symbol on drawings that are produced in one country for use among many countries, figure 22-3. The projection symbols are intended to promote the accurate exchange of technical information through drawings.

(A) FIRST ANGLE PROJECTION   (B) THIRD ANGLE PROJECTION

FIGURE 22-3 ISO projection symbols (enlarged).

The United States and Canada use the third-angle system of projection for drawings. Other countries, however, use a different system which is known as the first-angle projection system. The purpose of introducing the ISO projection symbols is to indicate that there is a continuously increasing international exchange of drawings for the production of interchangeable parts. Thus, the symbol tells whether the drawing follows the third-angle or the first-angle projection system.

The ISO projection symbol, the notation on tolerances, and information on whether metric and/or inch dimensions are used on the drawing should appear as notes either within the title block or adjacent to it, figure 22-4.

FIGURE 22-4 Dimension, tolerance, and symbol notes in a title block.

45° x 3.2

7.04 DIA DRILL, REAM TWO HOLES
$7.1437^{+0.0254}_{-0}$

$50.80\,\phi^{+0}_{-0.0254}$ Ⓓ

22.23

Ⓑ

203.2

139.7

6.35

Ⓕ
31.75 ϕ

19.05

$\frac{1}{4}$ – 20 UNC – 1B x 12.7 DEEP

Ⓔ $41.275\,\phi^{+0}_{-0.025}$

$1\frac{1}{2}$–12 UNF–LH–3A

$\frac{3}{8}$–16UNC–LH–
2B x 38.1 DEEP

44.45

38.1

114.3

Ⓒ

Ⓖ

Ⓗ

25.4 ϕ

31.75 ϕ

Ⓐ

NOTE:

ALL DIMENSIONS ARE GIVEN IN MILLIMETERS;
THREADS, IN THE UNIFIED THREAD SYSTEM.

SCALE:

0.75 mm = 1.00 mm

$\frac{3"}{4}$ = 1"

C.T. PUTNAM ASSOCIATES
WINSOR, CONN.

UNSPECIFIED TOLERANCES:	±0.4mm/±.016"	
DRAWN BY: T-P-O	FIRST ANGLE PROJECTION	
DATE: 10 • 12 • 82	PART: **POSITIONING SHAFT**	**BP-22**
PART NO. AA2176	MATL: C.D. STEEL	NO. REQD: 18

POSITIONING SHAFT (BP-22)

1. State what system of projection is used.

2. Indicate what unit of measurement is used for dimensioning.

3. Give the tolerance that applies on (a) metric dimensions and (b) dimensions converted to the inch system, when no tolerance is specified.

4. Compute the basic metric dimension for overall length Ⓐ rounded off to two decimal places.

5. Determine the basic center-to-center distance between the two reamed holes Ⓑ, in millimeters.

6. Find the length of the tapered portion Ⓒ.

7. Determine the equivalent customary (inch) units for each linear dimension, rounded off to two decimal places. Use a four-place millimeter/inch conversion chart, if available.

8. Convert each metric diametral dimension to its equivalent inch dimension, correct to three decimal places.

9. Give the letter size drill which corresponds to the drill size of the two reamed holes.

10. State what system of representation is used for the threaded sections.

11. Convert the depth of each tapped hole to its equivalent in the inch system.

12. Tell what each part of the three thread dimensions designates.

13. Give the minimum and maximum limits of diametral dimensions Ⓓ and Ⓔ in metric and inch dimensions, correct to three decimal places.

14. Determine the minimum and maximum limits of diametral dimensions Ⓕ, Ⓖ, and Ⓗ in metric and inch dimensions, rounded to two decimal places.

15. Determine the (a) taper per foot of the tapered portion, correct to two decimal places, and (b) convert the taper per foot to its millimeter equivalent.

16. Compute the lower and upper dimensional limits of each linear dimension in both the metric and inch systems. Round off the dimensions to two decimal places.

ASSIGNMENT UNIT 22

Student's Name _____

1. _____

2. _____

3. (a) _____
   (b) _____

4. Ⓐ _____

5. Ⓑ _____

6. Ⓒ _____

7.

Millimeters	Inch Equiv.
___ =	___
___ =	___
___ =	___
___ =	___
___ =	___

8.

Millimeters	Inch Equiv.
___ =	___
___ =	___

9. _____

10. _____

11. _____

12. _____

13.

	Millimeters	Inch Equiv.
Ⓓ Min.	___ =	___
Max.	___ =	___
Ⓔ Min.	___ =	___
Max.	___ =	___

14.

	Millimeters	Inch Equiv.
Ⓕ Min.	___ =	___
Max.	___ =	___
Ⓖ Min.	___ =	___
Max.	___ =	___
Ⓗ Min.	___ =	___
Max.	___ =	___

15. (a) _____
    (b) _____

16.

Dimension	Metric (mm)		Inch	
	Min.	Max.	Min.	Max.

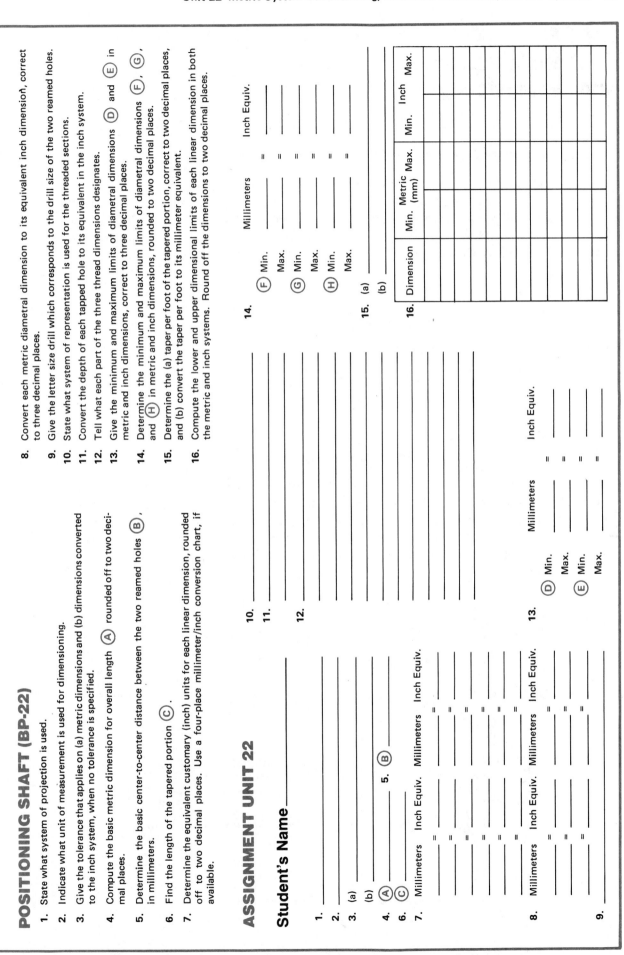

UNIT 23
First-angle Projection
and Dimensioning

PROJECTION BASED ON SYSTEM OF QUADRANTS

The position each view of an object occupies, as treated thus far in this text, is based upon the United States and Canadian standard of third-angle projection. Third-angle projection is derived from a theoretical division of all space into four quadrants. The horizontal plane in figure 23-1 represents an x-axis. The vertical plane is the y-axis. The four quadrants produced by the two planes are shown as I, II, III, and IV.

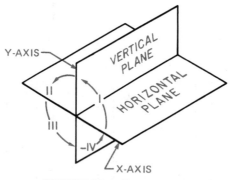

FIGURE 23-1 Four quadrants.

It is possible to place an object in any one of the four quadrants. Views of the object may then be projected. The third quadrant was adopted in the United States and Canada because the projected views of an object occupy a natural position. Drawings produced by such standards of projection are comparatively easy to interpret. Each view is projected so the object is represented as it is seen. The principles of third-angle projection are reviewed graphically in figure 23-2.

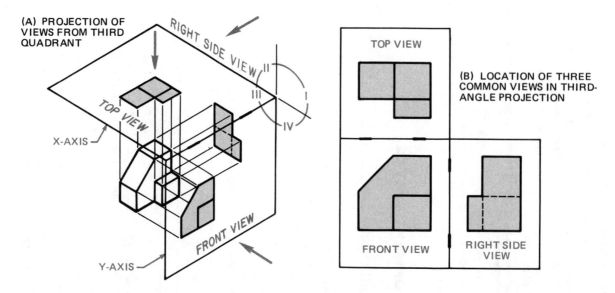

FIGURE 23-2 Third-angle projection.

METRIC: FIRST-ANGLE PROJECTION

Increasing worldwide commerce and the interchange of materials, instruments, machine tools, and other precision-made parts and mechanisms require the interpretation of drawings which have been prepared according to different systems of projection. The accepted standard of projection of the Common Market and other countries of the world relates to the first quadrant (I). The system, therefore, is identified as *first-angle projection.*

Dimensions given in the metric system are used with first-angle drawings. As world leaders in business and industry, the United States and Canada are also using a great number of first-angle projection drawings.

The quadrants and x and y axes in first-angle projection are shown in figure 23-3. Arrows are used in figure 23-3A to illustrate positions from which the object in quadrant I may be viewed. The three common views are named Front View, Top View, and Left-Side View. These views are positioned in figure 23-3B as they would appear (without the imaginary projection box outline) on a drawing.

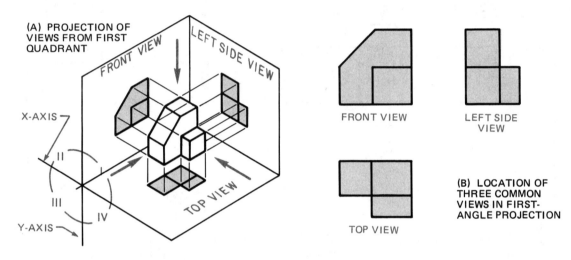

FIGURE 23-3 First-angle projection showing position of views.

In contrast with third-angle projection, the top view in first-angle projection appears under the front view. Similarly, the left-side view is drawn in the position occupied by the right-side-view in third-angle projection. This positioning, which is not natural, makes it more difficult to visualize and interpret first-angle drawings. Each view in first-angle projection is projected through the object from a surface of the object to the corresponding projection plane. However, the views in both first-angle and third-angle projection provide essential information for the craftsperson to produce or assemble a single part or a complete mechanism.

DIMENSIONING FIRST-ANGLE DRAWINGS

Dimensions on first-angle drawings are in metric units of measure. When a numerical value is less than 1, the first numeral is proceded by a zero and a decimal point. For example, the decimal .733 is dimensioned to read 0.733. Certain countries of Europe use a comma in the place of a decimal point. In this case the decimal 0.733 is dimensioned to read 0,733. Unless all measurements on the drawing are in millimeters, this value is followed by the unit of measurement, like mm, to read 0,733 mm.

NOTE: ALL DIMENSIONS IN MM

VIEW II

VIEW I

VIEW III

6.5 DRILL, 13.5 C'BORE,
9 DEEP, 3 HOLES

UNSPECIFIED TOLERANCES ± 0.4	LOCATING POINTS	
DRAWN: TPO	FIRST ∡ PROJECTION	DATE 11/10/82
CHK'D: SBO		
APPVD: JEF/CGW		

PUMP SYSTEMS MOUNT   **BP-23**

PART NO. 479 BB, 76	MAT'L STEEL FORGING KSN 208	NO. REQ'D 960

BUCHANAN
MOTOR WORKS
GLASGOW, SCOTLAND

PUMP SYSTEMS MOUNT (BP-23)

1. State what system of projection is used.

2. Name views I, II, and III.

3. Give the symbol which indicates a diameter.

4. Tell what the symbol △ indicates.

5. State what the comma denotes in the 36,6 dimension.

6. Indicate what tolerance to use if none is given.

7. Give the specifications for the counterbored holes (T).

8. Determine the upper and lower limit dimensions for diameters (A) and (B).

9. Give the ISO designation for threads (C). Describe what each value means.

10. Locate surfaces (D) and (E) in view III.

11. State the minimum diameter for hole (F).

12. Give the radius of fillet (G).

13. Compute maximum overall dimension (H).

14. Give the minimum width of undercut (I).

15. Locate surfaces (J) and (K) in view III and surface (L) in view I.

16. Compute maximum overall dimension (M), adding all the + tolerances to the basic dimensions.

17. Give minimum diameters (N) and (O).

18. State the thread size for (Q).

19. Locate surface (R) in view III.

20. Give the minimum diameter and width of holes (U) and (V).

21. Determine maximum overall dimensions (W) and (X).

22. Determine the basic dimension between surfaces (E) and (K).

ASSIGNMENT UNIT 23

Student's Name _____

1. _____

2. I = _____ II = _____
   III = _____

3. _____

4. _____

5. _____

6. _____

7. _____

8. Upper limit (A) = _____ (B) = _____
   Lower limit (A) = _____ (B) = _____

9. _____

10. (D) = _____ (E) = _____

11. (F) = _____

12. (G) = _____

13. (H) = _____

14. (I) = _____

15. (J) = _____ (K) = _____
    (L) = _____

16. (M) = _____

17. (N) = _____ (O) = _____

18. (Q) = _____

19. (R) = _____

20. (U) = _____

    (V) = _____

21. (W) = _____
    (X) = _____

22. Basic dimension = _____

UNIT 24
Dual Systems of Screw Threads, Dimensioning, and Tolerances

SCREW THREAD CONVERSION

A dual system of screw threads will exist for many years until the advantages of the *Unified Inch* system of America, Britain, and Canada are combined with those of the International Organization for Standardization of Metric Thread Series (ISO) to produce a single new International Standard Screw Thread System.

In the meantime, screw threads may be designated either by the present American Unified thread sizes or by SI Metric sizes. The metric thread series ranges from 0.30 UNM *(Unified Metric)* to 1.40 UNM, and from M1.6 (with a pitch of 0.35 millimeter) to M39 x 2. The M39 x 2 (SI designation) indicates that the thread has an outside diameter of 39 millimeters and a thread pitch of 2 millimeters.

Screw thread conversion tables usually contain engineering data on the outside diameter and pitch for SI designated numbers, the present American Unified thread sizes, and the best SI metric equivalent for certain American thread sizes.

For example, a 1/2-20 UNF designation on a drawing indicates that the outside diameter of the thread is .500" (for the 1/2") and the pitch is 20 threads per inch. Although there is no current precise metric equivalent, the outside diameter and pitch, expressed in metric units, is 12.7 mm O.D. and 1.27 mm pitch. The threads illustrated on the tap and by the drawing of a similarly threaded hole are dual dimensioned as shown in figure 24-1.

FIGURE 24-1 Dual dimensioning a tapped hole.

CONTROLLING DIMENSIONS

The usual designation for millimeter dimensions on a drawing is mm and for meters, m. If all dimensions are in the metric system, the drawing usually carries a note to indicate what unit of measurement or scale is used.

When a drawing contains dimensions in both the inch and the metric systems, the dimension in which the product was designed *(controlling dimension)* appears above the dimension line. The converted value appears below the line, figure 24-2.

FIGURE 24-2 Controlling dimensions.

104

DUAL DIMENSIONING: CUSTOMARY INCH AND SI METRIC

As industry changes gradually to include more SI metric dimensioning, it will be necessary to use *dual dimensioning*. When parts are dual-dimensioned, it means that the parts may be produced in any country regardless of whether the customary inch system dimensions or SI metric dimensions, or both, are used. Dual dimensioning implies that each dimension is given both in metric and English units. The metric dimension (including tolerances) is given on one side of the dimension line; the inch dimension is given on the opposite side of the line. For example, if dual dimensioning is used on a dimension of 83 millimeters (or 3.268″) the dimension is shown as,

If a tolerance of ±0.2 mm applies to the base dimension of 83 mm, the dimension and tolerance in the customary inch system equals 3.268″ ± .008″. The dual dimension is represented as,

Instead of a dimension line between the dimensions in the two systems, some industries use the slash (/) symbol between the metric and customary inch units for dual dimensioning.

$$\vdash\!\!-83\text{mm} \pm 0.2\text{mm}/3.268″ \pm.008″\!\!-\!\!\rightarrow\vdash$$

UNILATERAL AND EQUAL/UNEQUAL BILATERAL TOLERANCES

The previous examples showed equal tolerances in both directions (±). However, some tolerances may be unequal or unilateral. If unequal bilateral tolerances of $^{+0.2}_{-0.1}$ mm are applied to the 83 mm dimension, the drawing (with the scale and dimensional systems indicated) is dimensioned as,

NONSIGNIFICANT ZEROS

A *nonsignificant zero* refers to a zero that has no value in a number other than to simplify the reading of a dimension and to help in correctly interpreting the dimension. Drawings with dimensions in the customary inch system contain nonsignificant zeros to the right of the decimal point so that both the dimension and the tolerance have the same number of digits.

Example: 3.7222″ ± .0010″

By contrast, nonsignificant zeros are *not* added to a decimal value in the metric system. Thus, the number of digits in the base dimension and the tolerance may or may not be the same. For instance, the dimension and tolerance 72.5 mm ± 0.125 mm shows one digit in the fractional part of the base dimension (0.5 mm), and three digits (0.125 mm) for the bilateral decimal tolerance. A dimension that is less than one millimeter has a zero preceding the fractional value, for example 0.90 mm.

THIRD ANGLE PROJECTION:

SCALE: $\frac{5''}{16} = 1.0''$

DIMENSION: $\frac{INCH}{MILLIMETER}$

VIEW II

VIEW III

VIEW I

C.G. WHITEHURST CO.
KINGSTON, NY

UNSPECIFIED TOLERANCE: ±.016''/±0.40mm

FORGED ROCKER BRACKET   BP-24

DESIGN BY H.G.O.   PART: 027-3B   DATE: 9·26·82   MATL: STEEL FORGING   NO. REQD: 30

FORGED ROCKER BRACKET (BP-24)

1. Assume View I is the front view. Name Views II and III.

2. Name the dimensions above the line if these represent the product design dimensions.

3. State what dual dimensioning implies.

4. Determine what system of dimensioning is used.

5. Locate surfaces (A), (C), (D) and (E) in View III. Give the corresponding letter.

6. Give the letters in View II that identify surfaces (F), (G) and (H).

7. Locate surfaces (I), (J), (K), (L), (M), (N), (O) and (P) in View III.

8. Determine the distance between the center of the .750″ hole and surface (G) correct to two decimal places. Give the dimension in metric and customary inch units.

9. Give the center distance (F) in millimeters and inches, correct to two decimal places.

10. Locate thickness (I′) in both dimensioning systems, correct to one decimal place.

11. Indicate the tolerance to use if none is specified.

12. Interpret the meaning of the screw thread designation M38X4-LH-3A.

13. Give the outside diameter and pitch (in millimeters) for each metric screw thread.

14. Convert each metric screw thread to its equivalent American screw thread.

15. Give the upper and lower dimensional limits of diameters (J) and (J′) in inch units, to three decimal places.

16. Determine the minimum and maximum metric dimensions for (K′) and (L′), correct to three decimal places.

ASSIGNMENT UNIT 24

Student's Name _____

1. _____

2. _____

3. _____

4. _____

5. (A) = _____
   (C) = _____
   (D) = _____
   (E) = _____

6. (F) = _____
   (G) = _____
   (H) = _____

7. (I) = _____
   (J) = _____
   (K) = _____
   (L) = _____
   (M) = _____
   (N) = _____
   (O) = _____
   (P) = _____

8. _____

9. _____

10. _____

11. _____

12. _____

13. _____

14. _____

15. (J) = _____
    (J′) = _____

16. (K′) = _____
    (L′) = _____

UNIT 25
Cutting Planes, Full Sections, and Section Lining

An exterior view shows the object as it looks when seen from the outside. The inside details of such an object are shown on the drawing by hidden lines.

As the details inside the part become more complex, additional invisible lines are needed to show the hidden details accurately. This tends to make the drawing increasingly more difficult to interpret. One technique the draftsperson uses on such drawings to simplify them is to cut away a portion of the object. This exposes the inside surfaces. On cutaway sections, all of the edges that are visible are represented by visible edge or object lines.

CUTTING
PLANE
LINE

(A) IMAGINARY PLANE CUTTING THROUGH OBJECT

(B) CUTTING PLANE AND
FRONT PORTION OF OBJECT
REMOVED

FIGURE 25-1 Cutting plane and its application.

To obtain a sectional view, an imaginary cutting plane is passed through the object as shown in figure 25-1A. Figure 25-1B shows the front portion of the object removed. The direction and surface through which the cutting plane passes is represented on the drawing by a cutting plane line. The exposed surfaces, which have been theoretically cut through, are further identified by a number of fine slant lines called *section* or *cross-hatch* lines. If the cutting plane passes completely through the object, the sectional view is called a *full section.* A section view is often used in place of a regular view.

① ARROWS SHOW DIRECTION

② LETTERS IDENTIFY SECTION

FIGURE 25-2 The cutting plane line.

CUTTING PLANE LINES

The cutting plane line is either a thick (heavy) line with one long and two short dashes or a series of thick (heavy), equally spaced long dashes. Examples are shown in figure 25-2. The line represents the edge of the cutting plane. The arrowheads on the ends of the cutting plane line show the direction in which the section is viewed. A capital letter is usually placed at the end of the cutting plane line to identify the section.

SECTION A-A

(A) EXTERIOR VIEW   (B) FULL SECTION VIEW OF INTERIOR

FIGURE 25-3 Section views simplify internal details.

CROSS HATCHING OR SECTION LINING

The interpretation of a sectional drawing is simplified further by *cross hatch* or *section lines.* These uniformly spaced thin lines are drawn at a 45° angle as shown in figure 25-3. It is easy to distinguish one part from another on a cutaway section when section lines are used.

Section lines also serve the purpose of identifying the kind of material. Combinations of different lines and symbols are used for such metals as cast iron and steel and other groups of related materials. Section lining symbols of common materials are illustrated in figure 25-4.

CAST IRON

STEEL

BRASS, BRONZE, COPPER AND COMPOSITIONS

WHITE METAL, ZINC, LEAD, BABBITT AND ALLOYS

ALUMINUM AND MAGNESIUM

RUBBER, PLASTIC, AND ELECTRICAL INSULATION

WATER AND OTHER LIQUIDS

WOOD

FIGURE 25-4 Section linings identify materials.

ALL FILLETS AND
ROUNDS .25R

SECTION B-B

BODY

DRILL AND C'BORE FOR
3/8 ALLEN CAP SCREWS
4 HOLES

NO. 31
DRILL

20°

$\frac{3}{8} \times \frac{5}{32}$ KEYSEAT

SECTION A-A

MAT'L. C. I.	QTY. 125	ORDER NO. E S - 10 - 210				
DWG. BY A.V.P	CK. T.P.O.	APPVD. C.T.O.				

UNLESS OTHERWISE SPECIFIED
TOLERANCES ON DIMENSIONS ARE:
 	
    DECIMAL DIMENSIONS ± .01'' (TWO-PLACE)
                              ± .002'' (THREE-PLACE)
 	
    FRACTIONAL DIMENSIONS ± $\frac{1''}{64}$
 	
    ANGULAR DIMENSIONS ± 10'
			②	WAS 3.270	9-2-82	J.E.F.

			①	WAS 4$\frac{7}{8}$	9-2-82	J.E.F.
			NO.	CHANGE	DATE	BY
			IDLER SHAFT SUPPORT			**BP-25**

IDLER SHAFT SUPPORT (BP-25)

1. Give the specifications for the counter-bored holes.

2. How deep is the counterbored portion of the holes?

3. Give the specifications for the flat key-way.

4. What size are the fillets and rounds?

5. What size drill is used for the hole drilled at an angle?

6. Give the tolerance on the 20° dimension.

7. Compute angle (C) from dimensions given on the drawing.

8. Give the maximum diameter for the 1" hole.

9. What is the largest size to which diameter (B) can be bored?

10. How are the machined surfaces indicated?

11. What type line shows where Section A-A is taken? In what view?

12. Determine dimensions (A) and (D) .

13. Determine from Section B-B what material is required.

14. How wide and thick is the body?

15. Indicate what two changes were made from the original drawing.

16. Determine minimum dimension (E) .

17. Compute maximum dimension (F) .

18. What is upper limit of dimension (G) ?

19. Why is section A-A a full section?

20. Show the section linings for aluminum and for steel.

21. Identify (a) the system of dimensioning and (b) the classification of the unspecified tolerances.

22. Change the tolerances as follows: two-place decimal dimensions $^{+.02''}_{-.00''}$ three-place $^{+.003''}_{-.000''}$ Fractional $+\frac{1}{64}''$, $-0''$. Angular $^{+10'}_{-0'}$

    Then compute the upper and lower limit for angle (C) and dimensions (A) (D) (E) (F) and (G) .

ASSIGNMENT UNIT 25

Student's Name _____

1. _____
_____

2. _____

3. _____

4. _____ 5. _____

6. _____ 7. (C) = _____

8. _____ 9. _____

10. _____

11. _____

12. (A) = _____ (D) = _____

13. _____

14. _____

15. _____
_____

16. (E) = _____

17. (F) = _____

18. (G) = _____

19. _____
_____

20. Aluminum [   ] Steel [   ]

21. (a) _____

    (b) _____

22. Angle (C) = _____

    (A) = _____

    (D) = _____

    (E) = _____

    (F) = _____

    (G) = _____

UNIT 26
Half Sections, Partial Sections, and Conventional Breaks

HALF SECTIONS

The internal and external details of a part may be represented clearly by a sectional view called a *half section.* In a half-section view, one-half of the object is drawn in section and the other half is drawn as an exterior view, figure 26-1. The half section is used principally where both the inside and outside details are symmetrical and where a full section would omit some important detail in an exterior view.

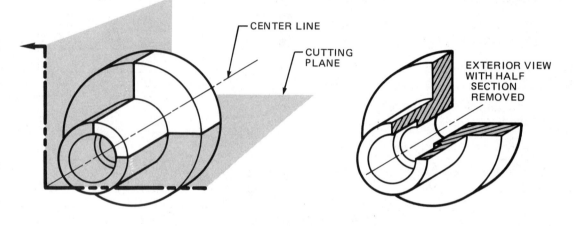

FIGURE 26-1 Half-section view.

Theoretically, the cutting plane for a half-section view extends halfway through the object, stopping at the axis or center line, figure 26-2. Drawings of simple symmetrical parts may not always include the cutting plane line or the arrows and letters showing the direction in which the section is taken. Also, hidden lines are not shown in the sectional view unless they are needed to give details of construction or for dimensioning.

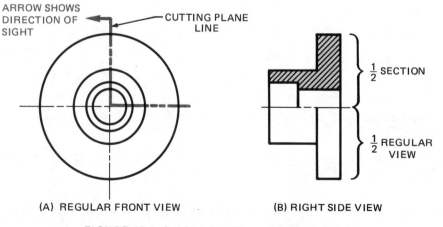

FIGURE 26-2 Cutting plane line and half section view.

112

BROKEN OR PARTIAL SECTIONS

On some parts, it is not necessary to use either a full or a half section to expose the interior details. In such cases, a *broken-out* or *partial section* may be used, figure 26-3. The cutting plane is imagined as being passed through a portion of the object and the part in front of the plane is then broken away. The break line is an irregular freehand line which separates the internal sectioned view and the external view.

FIGURE 26-3 Example of a broken-out or partial section.

CONVENTIONAL BREAKS

A long part with a uniform cross section may be drawn to fit on a standard size drawing sheet by cutting out a portion of the length. In this manner, a part may be drawn larger to bring out some complicated details.

The cutaway portion may be represented by a conventional symbol which does two things: (1) it indicates that a portion of uniform cross section is removed, and (2) it shows the internal shape. A few conventional symbols which are accepted as standard are illustrated in figure 26-4.

FIGURE 26-4 Standard break symbols.

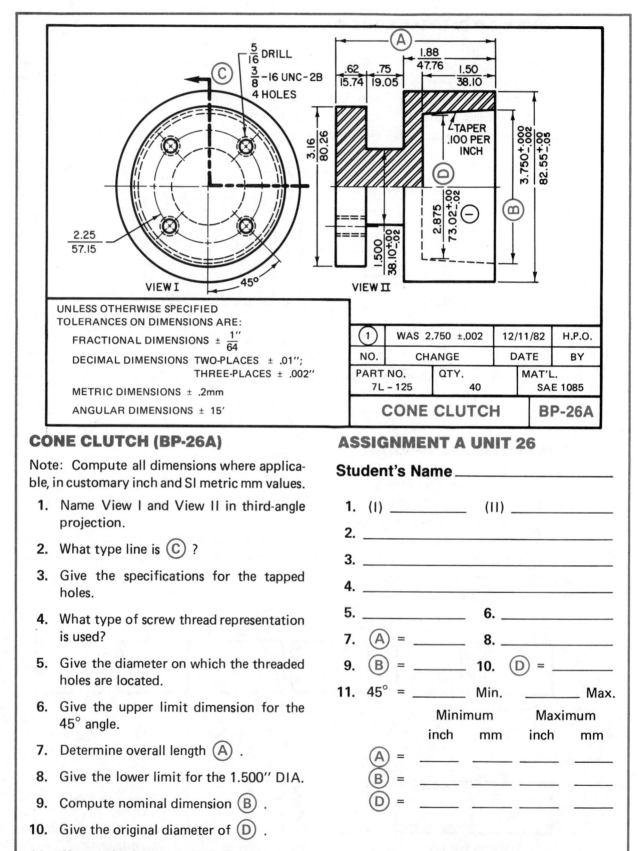

①	WAS 2.750 ±.002	12/11/82	H.P.O.
NO.	CHANGE	DATE	BY
PART NO. 7L – 125	QTY. 40	MAT'L. SAE 1085	

CONE CLUTCH   **BP-26A**

CONE CLUTCH (BP-26A)

Note: Compute all dimensions where applicable, in customary inch and SI metric mm values.

1. Name View I and View II in third-angle projection.

2. What type line is Ⓒ ?

3. Give the specifications for the tapped holes.

4. What type of screw thread representation is used?

5. Give the diameter on which the threaded holes are located.

6. Give the upper limit dimension for the 45° angle.

7. Determine overall length Ⓐ .

8. Give the lower limit for the 1.500" DIA.

9. Compute nominal dimension Ⓑ .

10. Give the original diameter of Ⓓ .

11. Change the tolerances as follows:
Two-place decimal dimensions $^{+.00}_{-.02}$; three-places, $^{+.000}_{-.0025}$"; millimeter dimensions $^{+.00}_{-.50}$; Angular dimensions $^{+5'}_{-15'}$. Then, compute the upper and lower limits for the 45° angle and dimensions Ⓐ , Ⓑ , and Ⓓ .

ASSIGNMENT A UNIT 26

Student's Name _____

1. (I) _____   (II) _____

2. _____

3. _____

4. _____

5. _____   6. _____

7. Ⓐ = _____   8. _____

9. Ⓑ = _____   10. Ⓓ = _____

11. 45° = _____ Min. _____ Max.

	Minimum		Maximum	
	inch	mm	inch	mm
Ⓐ =	_____	_____	_____	_____
Ⓑ =	_____	_____	_____	_____
Ⓓ =	_____	_____	_____	_____

ADAPTER BRACKET BP-26B

ADAPTER BRACKET (BP-26B)

Note: Compute all dimensions, where applicable, in customary inch and SI metric (mm) values.

1. Name View I and View II (third-angle projection).

2. How many outside machined surfaces are indicated in View II?

3. Give the upper and lower limits of the bored hole.

4. How deep is the hole bored?

5. What is the drill size for the tapped hole?

6. What size pipe tap is used?

7. Give the center-to-center distance of the elongated slot.

8. Compute dimension (A).

9. Determine the overall width of the pad.

10. Compute dimension (B).

11. Add a $^{+.01''}_{-.00}$ tolerance to all two-place decimal dimensions, and $^{+.20}_{-.00}$ mm to all metric dimensions. Then, determine the upper and lower limit dimensions for:
    (a) The depth of the bored hole.
    (b) The center-to-center distance between the elongated slot and the bored hole.

ASSIGNMENT B UNIT 26

Student's Name _____

1. (I) _____
   (II) _____

2. _____

3. Upper _____ '' _____ mm
   Lower _____ '' _____ mm

4. _____ '' _____ mm

5. _____

6. _____

7. _____ '' _____ mm

8. (A) = _____ '' _____ mm

9. _____

10. (B) = _____

	Minimum		Maximum	
	inch	mm	inch	mm
11. (a)				
(b)				
(c)				
(d)				
(e) (A)				
(B)				

(c) The overall width of the bracket (View I).
(d) The overall bracket length.
(e) Dimensions (A) and (B).

SECTION 6
Welding Drawings

UNIT 27
Symbols, Representation, and Dimensioning

WELDED STRUCTURES AND SYMBOLS

Many parts which were formerly cast in foundries are now being constructed by welding. Complete welding and machining information is conveyed from the part designer to the welder, machinist, and machine technician by graphic symbols. The symbols indicate the required type of weld, specific welding and machining dimensions, and other specifications.

TYPES OF WELDED JOINTS

There are five basic types of welded joints specified on drawings. Each type of joint is identified by the position of the parts to be joined together. Parts which are welded by using butt, corner, tee, lap, or edge-type joints are illustrated in figure 27-1.

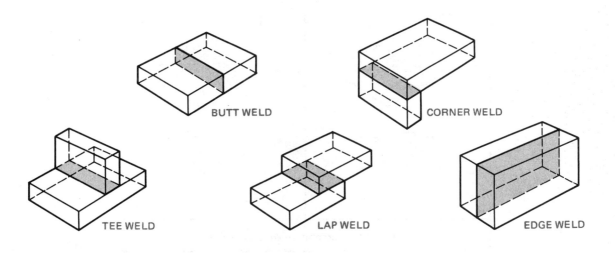

FIGURE 27-1 Basic types of welded joints.

ANSI SYMBOLS FOR ARC AND GAS WELDS

The most commonly used arc and gas welds for fusing parts are shown in figure 27-2. The welds are grouped as follows: (1) bead, (2) fillet, (3) plug and slot, and (4) groove. The groove welds are further identified as (A) square, (B) vee, (C) bevel, (D) U, and (E) J.

117

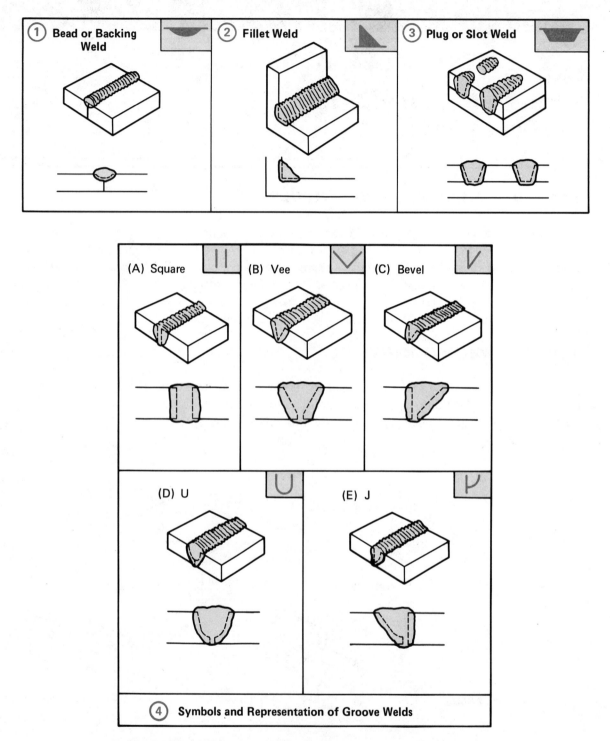

FIGURE 27-2 ANSI symbols and examples of common arc and gas welds.

RESISTANCE WELDING: TYPES AND SYMBOLS

Resistance welding is another method of fusing parts. The fusing temperature is produced in the particular area to be welded by applying force and passing electric current between two electrodes and the parts. Resistance welding does not require filler metal or fluxes. The symbols for general types of resistance welds are given in figure 27-3.

Types of Resistance Welds			
Spot	Projection	Seam	Flash or Upset

FIGURE 27-3 Symbols for resistance welds.

SUPPLEMENTARY WELD SYMBOLS

General supplementary weld symbols are shown in figure 27-4. These symbols convey additional information about the extent of welding, location, and contour of the weld bead. The contour symbols are placed above (OTHER SIDE) or below (ARROW SIDE) the weld symbol.

Field Weld	Weld All Around	Melt Through	Contour		
			Flush	Convex	Concave

FIGURE 27-4 Supplementary weld symbols.

ANSI (AMERICAN WELDING SOCIETY) STANDARDS FOR SYMBOLS, REPRESENTATION ON DRAWINGS, AND DIMENSIONING

The main parts of the ANSI standard welding symbol (adopted in collaboration with the American Welding Society) are identified in figure 27-5.

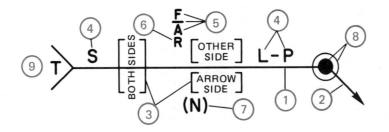

FIGURE 27-5 Parts of an ANSI standard welding symbol.

The symbols used for parts ① through ⑨ are given in column (A), figure 27-6. A brief description of the function of each part follows in column (B). Examples of applications appear in column (C). ANSI standards are used on many metric drawings, pending the establishment of SI metric standards.

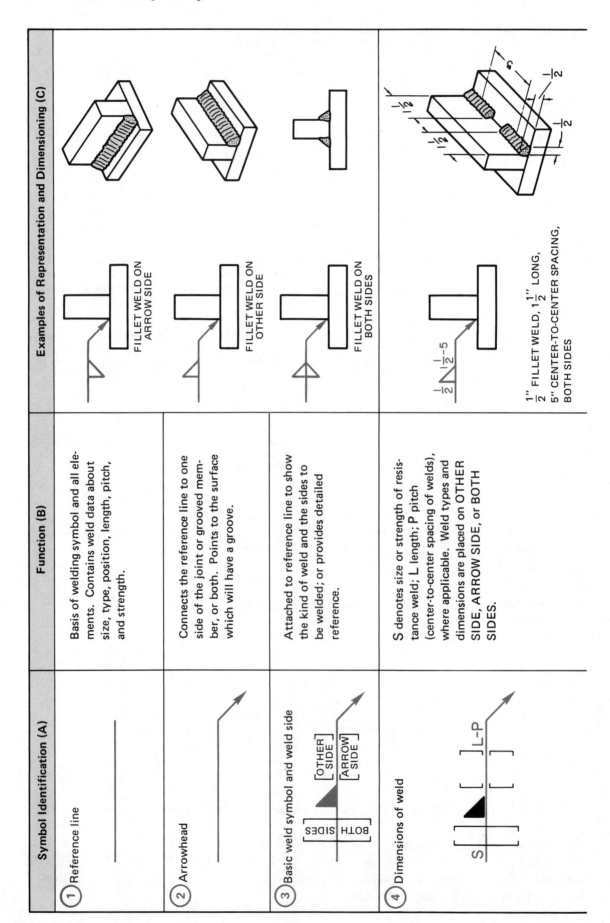

Symbol Identification (A)	Function (B)	Examples of Representation and Dimensioning (C)
① Reference line	Basis of welding symbol and all elements. Contains weld data about size, type, position, length, pitch, and strength.	FILLET WELD ON ARROW SIDE FILLET WELD ON OTHER SIDE
② Arrowhead	Connects the reference line to one side of the joint or grooved member, or both. Points to the surface which will have a groove.	FILLET WELD ON BOTH SIDES
③ Basic weld symbol and weld side	Attached to reference line to show the kind of weld and the sides to be welded; or provides detailed reference.	
④ Dimensions of weld	S denotes size or strength of resistance weld; L length; P pitch (center-to-center spacing of welds), where applicable. Weld types and dimensions are placed on OTHER SIDE, ARROW SIDE, or BOTH SIDES.	1" FILLET WELD, 1½" LONG, 5" CENTER-TO-CENTER SPACING, BOTH SIDES

⑤	Finish, contour, and groove angle F A R	F specifies the finish of the weld: chip (C), machine (M), and grind (G). ⌒ shows the contour of the weld: flush ▬, convex ⌒, and concave ⌣. A indicates the groove angle; included angle of countersink for plugs.	G GROUND WELD ⌒ CONVEX FORM ◣ FILLET WELD OTHER SIDE
⑥	Root opening R	R specifies the depth of filling for plug or slot welds.	U GROOVE, ARROW SIDE 12 mm DEEP, 30° ANGLE, 0 mm ROOT OPENING
⑦	Number of spots (N)	Number in parentheses gives the number of spots or welds.	✳ SPOT WELD, 8 mm DIAMETER (3) WELDS, 38 mm BETWEEN WELDS
⑧	Field weld and all around weld symbols	● indicates parts are to be welded in final assembly or in the field. ◯ signifies weld extends around the part.	↰FW FLASH WELD ✛ RESISTANCE WELD ◉ WELDED AROUND THE PART IN THE FIELD G FLAT GROUND FINISHED SURFACE
⑨	Tail T	T⟩ provides additional welding process information or specifications which are not otherwise shown by symbols.	

FIGURE 27-6 ANSI welding symbols, functions, and drawing representation

WELDED ROTARY PLUG (BP-27)

NOTE: All dimensions are to be given in both inch and millimeter values.

1. Name the two views.

2. List (a) the wall thickness of the tubing and (b) the material for the tube body and other parts.

3. Give the nomimal diameters of Ⓐ , Ⓑ , and Ⓒ .

4. Determine the maximum and minimum diameters for the bored hole Ⓓ .

5. Establish the maximum overall height Ⓔ and length Ⓕ .

6. Give the specifications for the reamed holes.

7. Identify (a) the minimum diameter and (b) the depth of the counterbored holes.

8. Specify the size of the welded pad.

9. Determine how many surfaces are to be machined.

10. Give the maximum dimensions of the circular end plate.

11. Indicate (a) the type of welded joint between the pad and the 2.00″ thick plate, and (b) give the weld specifications.

12. Interpret the weld specifications for the tubing and square end plate.

13. Draw the ANSI symbols which describe the weld for the circular end plate and tubing body.

14. Interpret the symbols and dimensions for the circular end plate and body weld.

ASSIGNMENT UNIT 27

Student's Name

1. View I _____ View II _____

2. (a) Wall thickness. _____
   (b) Materials. _____

3. Ⓐ _____ nominal diameter
   Ⓑ _____ nominal diameter
   Ⓒ _____ nominal diameter

4. Ⓓ Maximum diameter. _____
   Minimum diameter. _____

5. Ⓔ Height _____ Ⓕ Length _____

6. _____

7. (a) Minimum diameter _____
   (b) Minimum depth _____

8. _____

9. _____

10. _____

11. (a) Type. _____
    (b) Specifications. _____

12. _____

13. _____

14. _____

UNIT 28
Datums; Ordinate and Tabular Dimensioning

NUMERICAL CONTROL FUNCTIONS

Numerical control (NC) is defined as the operation of a machine by a series of coded instructions. The instructions comprise a system of numbers and letters. These are used in combination with a comma, dash, asterisk (*), ampersand (&), and other symbols.

NC instructions (commands) control the sequence of events for:

- *Positive positioning* the spindle in relation to horizontal (X), transverse (Y), vertical (Z), and other axes of the workpiece or machine tool.

- *Auxiliary functions* such as speed control and direction, feeds, and tool selection for machining stations.

- *On-Off* coolant flow and other *tooling functions.*

The logical organization *(documentation)* of commands is called a workpiece *program* or *NC programming.* Basic NC machine tools are programmed manually. However, control of machine tooling and processes may be extended by other *computer assisted machining (CAM)* systems.

NC programming is based on information provided on a drawing. The type drawing which is used requires fixed reference points for dimensioning and a rectangular coordinate system for uniformly establishing command information.

DATUMS

A *datum* is a reference feature from which dimensions are located and essential information is derived. A datum may be a point, line, plane, cylinder, or other exact feature. Datums are used in design, manufacturing, numerical control, and many other processes. Datums give location dimensions, furnish data for computations, and provide a reference source.

When a datum is specified, such as the two that are illustrated in figure 28-1, all features must be stated in relation to the datum reference. Different features of the part are located from the datum, not from other features. Datums must be clearly identified or easily recognizable. At least two, and often three, datums must be used to define or measure a part. Since all measurements are taken from datums (similar to those earlier used in base line dimensioning), errors in dimensioning are not cumulative.

ORDINATE DIMENSIONING AND NUMERICAL CONTROL MACHINE DRAWINGS

Ordinate dimensioning is a type of rectangular datum dimensioning. In ordinate dimensioning the dimensions are measured from two or three mutually related datum planes, figure 28-1. The datum planes are indicated as *zero coordinates.* Dimensions from the zero coordinates are represented on drawings as extension lines, without the use of dimension lines or arrowheads.

Specific features are located by the intersection of datum dimensions. Ordinate dimensioning is used when there would otherwise be a large number of dimensions and features and close tolerances are required. Datums and ordinate dimensioning are the foundation of numerically-controlled machine drawings and precision parts.

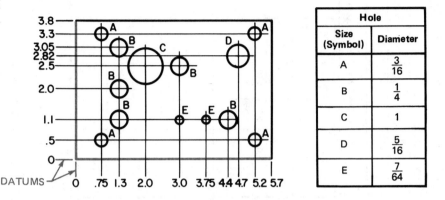

Hole	
Size (Symbol)	Diameter
A	$\frac{3}{16}$
B	$\frac{1}{4}$
C	1
D	$\frac{5}{16}$
E	$\frac{7}{64}$

FIGURE 28-1 Ordinate dimensioning.

TABULAR DIMENSIONING

Tabular dimensioning is another form of rectangular datum dimensioning. As the term implies, dimensions from intersecting datum planes are given in a table, figure 28-2. Tabular dimensioning helps to eliminate possible errors which could result from incorrectly reading a dimension when a large number are included on a drawing.

Values of x and y are measured along the respective datum lines from their intersection. The intersection is the origin of the datums. Tabular dimensioning is recommended where there are a great many repetitive features. These would make a dimensioned drawing difficult to read. Dimensions may also be given on a second or additional view for operations to be performed along Z or other axes.

Hole		Location (Axes)	
Identification	Size	X →	Y ↑
A₁	3/16	.75	3.3
A₂	3/16	5.2	3.3
A₃	3/16	.75	.5
A₄	3/16	5.2	.5
B₁	1/4	1.3	3.05
B₂	1/4	3.0	2.5
B₃	1/4	1.3	2.0
B₄	1/4	1.3	1.1
B₅	1/4	4.4	1.1
C	1	2.0	2.5
D	5/16	4.7	2.82
E₁	7/64	3.0	1.1
E₂	7/64	3.75	1.1

FIGURE 28-2 Coordinate chart and tabular dimension drawing.

COORDINATE CHARTS

The dimensions of a coordinate and features that are to be fabricated on coordinates are given in a coordinate chart, figure 28-2. Only those dimensions that originate at the datums are found in such a chart. The coordinates may specify sufficient information to fabricate the entire feature or only a portion of it.

Parts or components may be represented on drawings by ordinate or tabular dimensioning in the English or metric system of measurement or in both systems.

Hole	No. of Holes	Specifications
A	2	.1065 $+ .003$ / $- .000$ Dia 12 UNF-2B-1/2 deep
B	2	.125 $+ .000$ / $- .001$ Dia
C	4	.161 $+ .003$ / $- .000$ Dia – Tap 10-32
D	3	.1695 $+ .003$ / $- .000$ Dia
E	3	.562 $+ .002$ / $- .000$ Dia CSK 82° x .593 $+ .015$ / $- .000$ Dia

Hole/ Number	Distance X →	Distance Y ↑	Distance Z+
A1		2.563	
A2		5.188	
B1	1.157 ±.010	.063 ±.010	
B2	1.157 ±.010	2.844 ±.010	
C1	.289	1.584	
C2	.652	2.844	
C3	2.157	2.250	
C4	2.157	3.282	
D1	1.126	1.282	
D2	.595	1.282	
D3	1.126	2.250	
D4			P
D5	.595	5.272	R
D6	.595		T
E1			U
E2	1.407		W
E3			

CHANGES	DATE	NATURE
NOTES		
DATE 10/10/82	CKD BY CGW	ORDER NO. 12401

UNSPECIFIED TOLERANCES
FRACTIONS ± $\frac{1}{64}$ ″   DECIMALS ± .003″

DWG BY AEO

MODEL ALQ81675

FINISH PENETRATE OR EQUIVALENT

STOCK PER 100 PCS   114 LBS   60 FT

FORWARDS MACHINE WORKS

MATL CDS-SAE 1020

STACKER PLATE

BP-28

STACKER PLATE (BP-28)

1. Give the specification of the material in the Stacker Plates.

2. Determine the weight of 50 Stacker Plates.

3. Give the length of stock needed for 50 Stacker Plates.

4. State what modified systems of dimensioning are used.

5. Compute the maximum overall height and give the maximum thickness.

6. Determine the correct letter, designation or number for 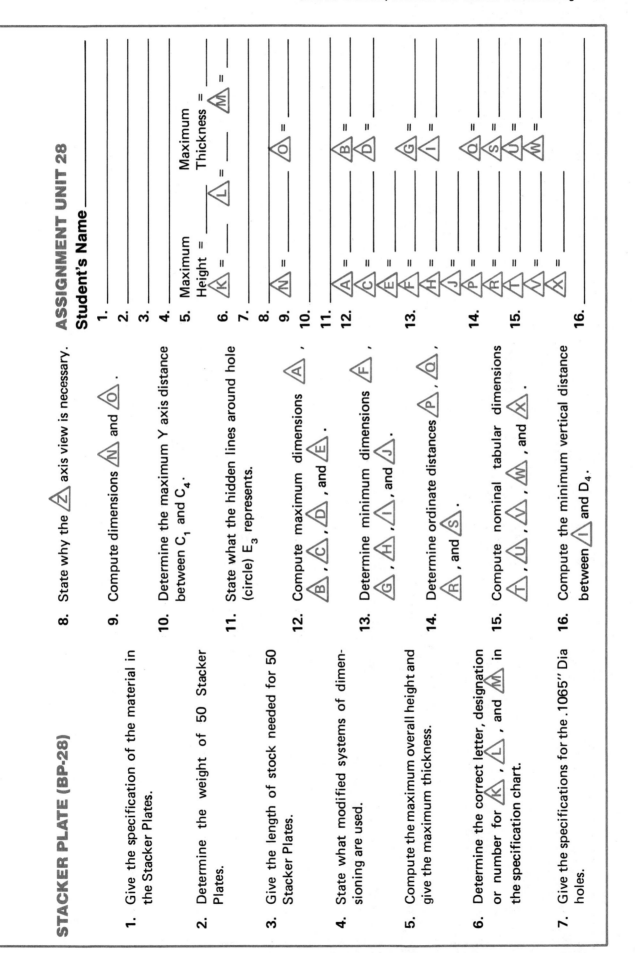K , L , and M in the specification chart.

7. Give the specifications for the .1065" Dia holes.

8. State why the Z axis view is necessary.

9. Compute dimensions N and O .

10. Determine the maximum Y axis distance between C_1 and C_4.

11. State what the hidden lines around hole (circle) E_3 represents.

12. Compute maximum dimensions A , B , C , D , and E .

13. Determine minimum dimensions F , G , H , I , and J .

14. Determine ordinate distances P , Q , R , and S .

15. Compute nominal tabular dimensions T , U , V , W , and X .

16. Compute the minimum vertical distance between I and D_4.

ASSIGNMENT UNIT 28

Student's Name

1. _____

2. _____

3. _____

4. _____

5. Maximum   Maximum
   Height = _____   Thickness = _____

6. K = _____   L = _____   M = _____

7. _____

8. _____

9. N = _____

10. O = _____

11. _____

12. A = _____

    C = _____

    E = _____

    B = _____

    D = _____

13. F = _____

    H = _____

    J = _____

    G = _____

    I = _____

14. P = _____

    R = _____

    Q = _____

    S = _____

15. T = _____

    V = _____

    U = _____

    W = _____

    X = _____

16. _____

PART 2

UNIT 29
Sketching Horizontal, Vertical, and Slant Lines

THE VALUE OF SKETCHING

Many parts or assembled units may be described clearly and adequately by one or more freehand sketches. Sketching is another way of conveying ideas rather than a method of making perfect completed drawings.

Technicians, therefore, in addition to being able to interpret drawings and blueprints accurately and easily, are often required to make sketches. By sketching additional views not found on blueprints, a craftsperson can study the part thoroughly to understand the process required to fabricate or machine it. Since the only tools needed are a soft pencil and any available paper, sketches may be made conveniently at any time or place.

SKETCHING LINES FREEHAND

The development of correct skills in sketching lines freehand is more essential for the beginner than speed. After the basic principles of sketching are learned and skill is acquired in making neat and accurate sketches, then stress should be placed on speed.

Although shop sketches are made on the job and special pencils are not required, a soft lead pencil with a cone-shaped point will produce the best results. For fine lines, use a fairly sharp point, for heavier lines, round the point more, figure 29-1.

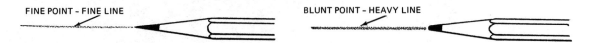

FINE POINT – FINE LINE

BLUNT POINT – HEAVY LINE

FIGURE 29-1 Sharpness of pencil point influences line weight.

SKETCHING HORIZONTAL LINES

The same principles of drafting which apply to the making of a mechanical drawing and the interpretation of blueprints are used for making sketches.

In sketching, the pencil is held 3/4" to 1" from the point so the lines to be drawn may be seen easily and there is a free and easy movement of the pencil. A free arm movement makes it possible to sketch smooth, neat lines as compared with the rough and inaccurate lines produced by a finger and wrist movement.

In planning the first horizontal line on a sketch, place two points on the paper to mark the beginning and end of the line, figure 29-2A. Place the pencil point on the first dot. Then, with a free arm movement and with the eyes focused on the right point, draw the line from left to right, figure 29-2B.

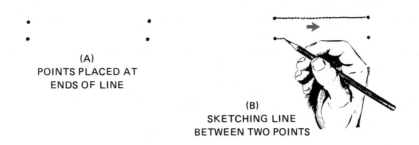

(A)
POINTS PLACED AT
ENDS OF LINE

(B)
SKETCHING LINE
BETWEEN TWO POINTS

FIGURE 29-2 Sketching horizontal lines.

Examine the line for straightness, smoothness, and weight. If the line is too light, a softer pencil or a more rounded point may be needed. On long lines, extra dots are often placed between the start and finish points of the line. These intermediate dots are used as a guide for drawing long straight lines.

SKETCHING VERTICAL LINES

The same techniques for holding the pencil and using a free arm movement apply in sketching vertical lines. Dots again may be used to indicate the beginning and end of the vertical line. Start the line from the top and move the pencil downward as shown in figure 29-3. For long lines, if possible, the paper may be turned to a convenient position.

FIGURE 29-3 Sketching vertical lines.

FIGURE 29-4 Sketching slant lines.

SKETCHING SLANT LINES

The three types of straight lines which are widely used in drawing and sketching are: (1) horizontal, (2) vertical, and (3) inclined or slant lines.

The slant line may be drawn either from the top down or from the bottom up. The use of dots at the starting and stopping points for slant lines is helpful to the beginner. Slant lines, figure 29-4, are produced with the same free arm movement used for horizontal and vertical lines.

USING GRAPH PAPER

Many industries recommend the use of graph paper for making sketches. Graph paper is marked to show how many squares are included in each inch. The combination of light ruled lines at a fixed number per inch makes it possibe to draw neat sketches which are fairly accurate in size. Throughout this text, squared or isometric graph sections are included for each sketching assignment. Other ruled section papers are available with diametric, oblique or perspective lines.

PUNCH PLATE   BP-29A

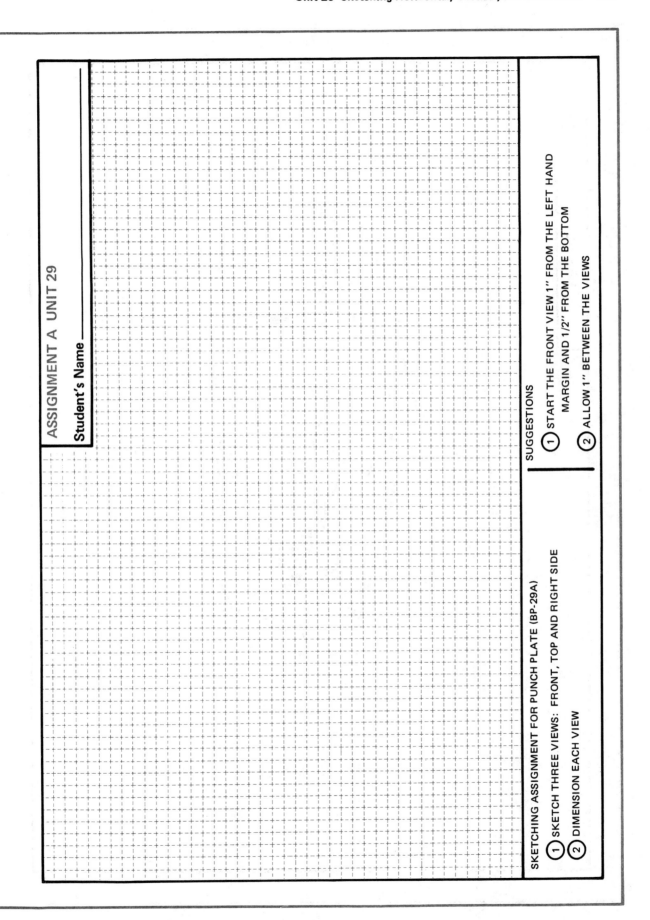

ASSIGNMENT A   UNIT 29

Student's Name _____

SUGGESTIONS

① START THE FRONT VIEW 1" FROM THE LEFT HAND
    MARGIN AND 1/2" FROM THE BOTTOM

② ALLOW 1" BETWEEN THE VIEWS

SKETCHING ASSIGNMENT FOR PUNCH PLATE (BP-29A)

① SKETCH THREE VIEWS: FRONT, TOP AND RIGHT SIDE

② DIMENSION EACH VIEW

BP-29B

TEE SLIDE

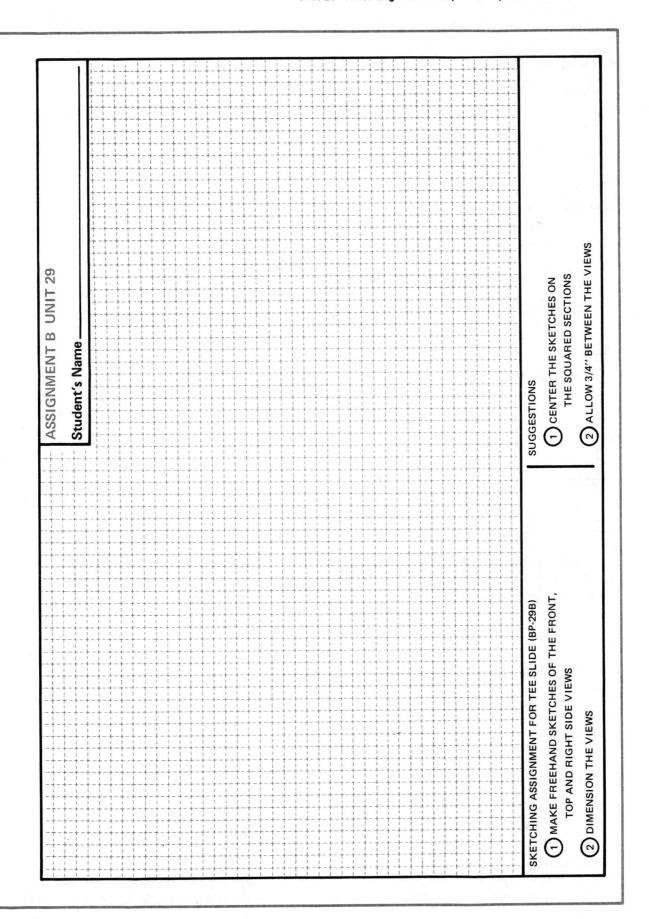

ASSIGNMENT B UNIT 29

Student's Name _____

SKETCHING ASSIGNMENT FOR TEE SLIDE (BP-29B)

1 MAKE FREEHAND SKETCHES OF THE FRONT, TOP AND RIGHT SIDE VIEWS

2 DIMENSION THE VIEWS

SUGGESTIONS

1 CENTER THE SKETCHES ON THE SQUARED SECTIONS

2 ALLOW 3/4'' BETWEEN THE VIEWS

UNIT 30
Sketching Curved Lines and Circles

ARCS CONNECTING STRAIGHT LINES

Often, when describing a part, it is necessary to draw both straight and curved lines. When part of a circle is shown, the curved line is usually called an *arc.*

When an arc must be drawn so that it connects two straight lines (in other words, is tangent to them), it may be sketched easily if five basic steps are followed, figure 30-1.

STEP 1 ▶ Extend the straight lines so they intersect.

STEP 2 ▶ Step off the same distance from the intersection (center) of both lines.

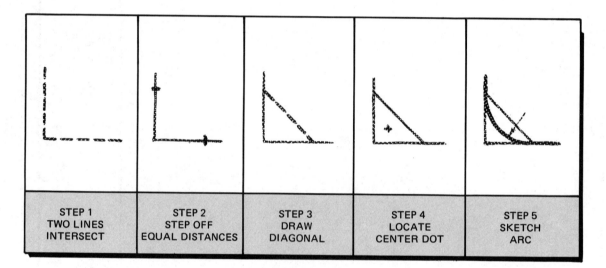

| STEP 1 TWO LINES INTERSECT | STEP 2 STEP OFF EQUAL DISTANCES | STEP 3 DRAW DIAGONAL | STEP 4 LOCATE CENTER DOT | STEP 5 SKETCH ARC |

FIGURE 30-1 Sketching an arc freehand.

STEP 3 ▶ Draw the diagonal line through these two points to form a triangle.

STEP 4 ▶ Place a dot in the center of the triangle.

STEP 5 ▶ Start at one of the lines and sketch an arc which runs through the dot and ends on the other line. Darken the arc and erase all unnecessary lines.

ARCS CONNECTING STRAIGHT AND CURVED LINES

Two other line combinations using straight lines and arcs are also very common. The first is the case of an arc connecting a straight line with another arc; the second is where an arc connects two other arcs. The same five basic steps are used as illustrated in figure 30-2.

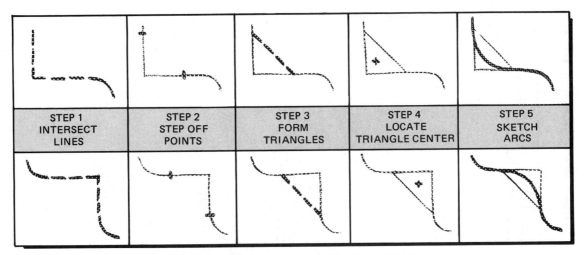

FIGURE 30-2 **Sketching arcs in combination with other arcs.**

SKETCHING CIRCLES

Shop sketches also require the drawing of circles or parts of a circle. While there are many ways to draw a circle freehand, a well-formed circle can be sketched by following five simple steps which are described and illustrated in figure 30-3.

STEP 1 ▶ Lay out the vertical and horizontal center lines in the correct location. Measure off half the diameter of the circle on each side of the two center lines.

STEP 1 LAYING OUT DIAMETER	STEP 2 DRAWING SQUARE	STEP 3 DRAWING DIAGONALS	STEP 4 LOCATING TRIANGLE CENTER	STEP 5 SKETCH ARCS

FIGURE 30-3 **Sketching a circle.**

STEP 2 ▶ Draw lightly two vertical and two horizontal lines passing through the points marked off on the center lines. These lines, properly drawn, will be parallel and will form a square.

STEP 3 ▶ Draw diagonal lines from each corner of the square to form four triangles.

STEP 4 ▶ Place a dot in the center of each triangle through which an arc is to pass.

STEP 5 ▶ Sketch the arc for one quarter of the circle. Start at one center line and draw the arc through the dot to the next center line. Continue until the circle is complete. Darken the circle and erase all the guide lines to simplify the reading of the sketch.

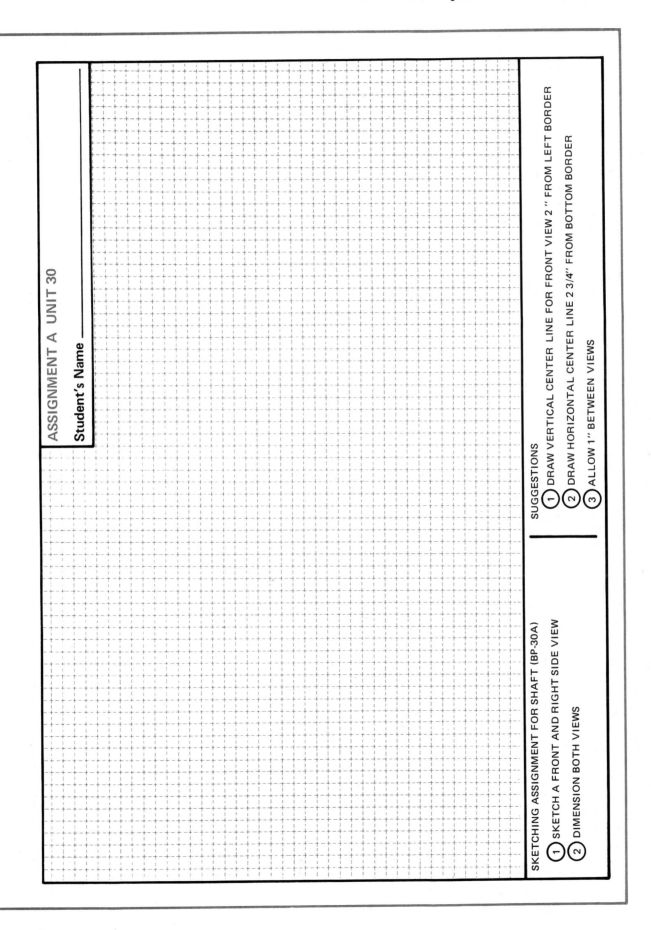

ASSIGNMENT A UNIT 30

Student's Name

SUGGESTIONS

① DRAW VERTICAL CENTER LINE FOR FRONT VIEW 2 " FROM LEFT BORDER

② DRAW HORIZONTAL CENTER LINE 2 3/4" FROM BOTTOM BORDER

③ ALLOW 1" BETWEEN VIEWS

SKETCHING ASSIGNMENT FOR SHAFT (BP-30A)

① SKETCH A FRONT AND RIGHT SIDE VIEW

② DIMENSION BOTH VIEWS

SLIDE BLOCK

BP-30B

$1\frac{1}{2}$

$3\frac{3}{4}$

$2\frac{1}{4}$

$\frac{3}{4}$

$1\frac{1}{8}$

$1\frac{1}{8}$

$2\frac{1}{4}$

2

$\frac{3}{4}$ DIA THROUGH HOLE

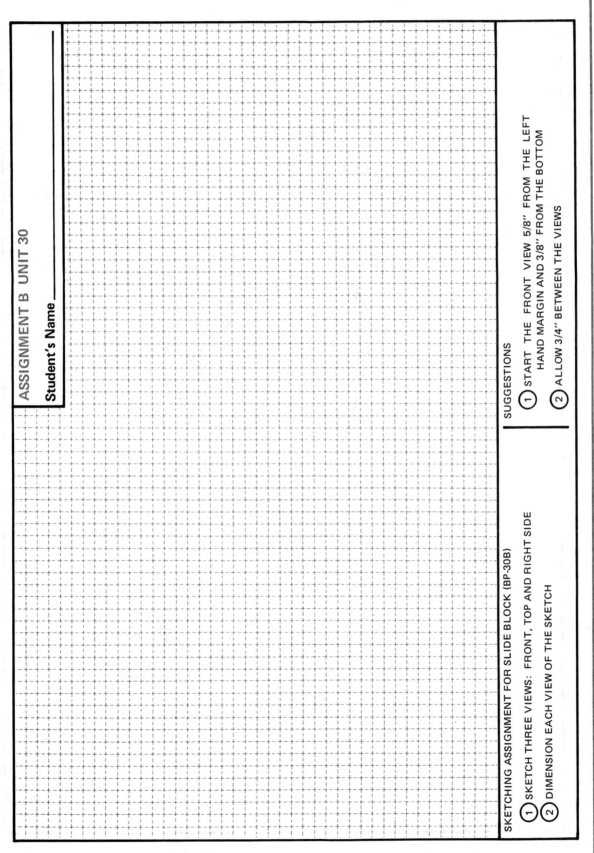

ASSIGNMENT B UNIT 30

Student's Name

SKETCHING ASSIGNMENT FOR SLIDE BLOCK (BP-30B)

① SKETCH THREE VIEWS: FRONT, TOP AND RIGHT SIDE

② DIMENSION EACH VIEW OF THE SKETCH

SUGGESTIONS

① START THE FRONT VIEW 5/8" FROM THE LEFT HAND MARGIN AND 3/8" FROM THE BOTTOM

② ALLOW 3/4" BETWEEN THE VIEWS

BP-30C

FLANGE

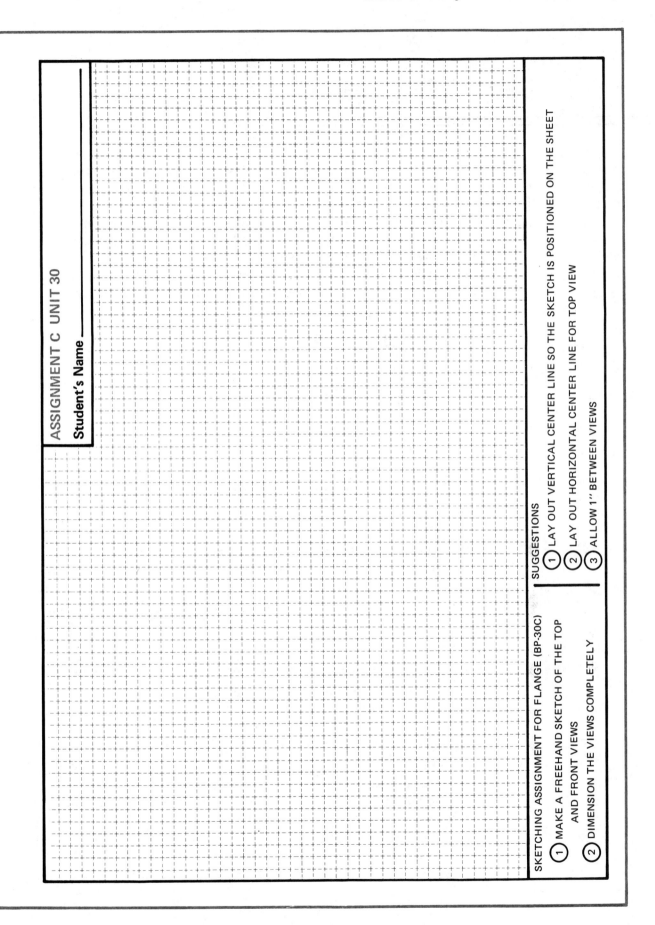

ASSIGNMENT C UNIT 30

Student's Name

SUGGESTIONS

① LAY OUT VERTICAL CENTER LINE SO THE SKETCH IS POSITIONED ON THE SHEET

② LAY OUT HORIZONTAL CENTER LINE FOR TOP VIEW

③ ALLOW 1" BETWEEN VIEWS

SKETCHING ASSIGNMENT FOR FLANGE (BP-30C)

① MAKE A FREEHAND SKETCH OF THE TOP AND FRONT VIEWS

② DIMENSION THE VIEWS COMPLETELY

UNIT 31
Sketching Irregular Shapes

Parts that are irregular in shape often look complicated to sketch. However, the object may be drawn easily if it is first visualized as a series of square and rectangular blocks. Then, by using straight, slant, and curved lines in combination with each other, it is possible to draw in the squares and rectangles the exact shape of the part.

The Shaft Support shown in figure 31-1 is an example of a machine part that can be sketched easily by *blocking*. The support must first be thought of in terms of basic squares and rectangles. After these squares and rectangles are determined, the step-by-step procedures which are commonly used to simplify the making of a sketch of an irregularly-shaped piece are applied. These step-by-step procedures are described next and their application is shown in figure 31-2.

FIGURE 31-1 Example of irregularly-shaped part to be sketched.

STEP 1 ▶   Lay out the two basic rectangles required for the shaft support. Use light lines and draw the outline to the desired overall sketch size.

STEP 2 ▶   Place a dot at the center of the top horizontal line and draw the vertical center line. Also, draw two vertical lines for the sides of the top rectangle.

STEP 3 ▶   Locate dots and draw horizontal lines from the left vertical line to locate:

    (A) The rectangle for the top of the support, and
    (B) The center line of the hole.

STEP 4 ▶   Draw the squares for the rounded corners of the base, top, and the center hole. In these squares:

    (A) Draw the diagonals and form the triangles, and
    (B) Locate dots in the centers of the triangles through which each arc will pass.

STEP 5 ▶ Draw the arcs and circle. Start at one diagonal and draw the arc through the dot to the next diagonal. Continue until the circle is completed.

STEP 6 ▶ Draw the side view if necessary.

STEP 7 ▶ Darken all object lines and dimension. Erase those lines used in construction that either do not simplify the sketch or are not required to interpret the sketch quickly and accurately.

FIGURE 31-2 Sketching an irregular shape.

The techniques described in this unit have been found by tested experience to be essential in training the beginner to sketch accurately.

As skill is developed in drawing straight, slant and curved lines freehand in combination with each other, some of the steps in sketching irregularly-shaped parts may be omitted and the process shortened.

TOOL BLOCK | BP-31

.625 REAM
$1\frac{1}{4}$ DIA SPOTFACE TO $\frac{1}{8}$ DEEP

FRONT VIEW

MATERIAL SAE 1080

$1R$

TOP VIEW

$\frac{5}{8}$

A

A

$\frac{3}{4}$

$\frac{5}{8}$

$1\frac{1}{2}$

$2\frac{3}{4}$

$\frac{3}{8}$

$\frac{1}{2}$

$\frac{1}{2}R$

1

2

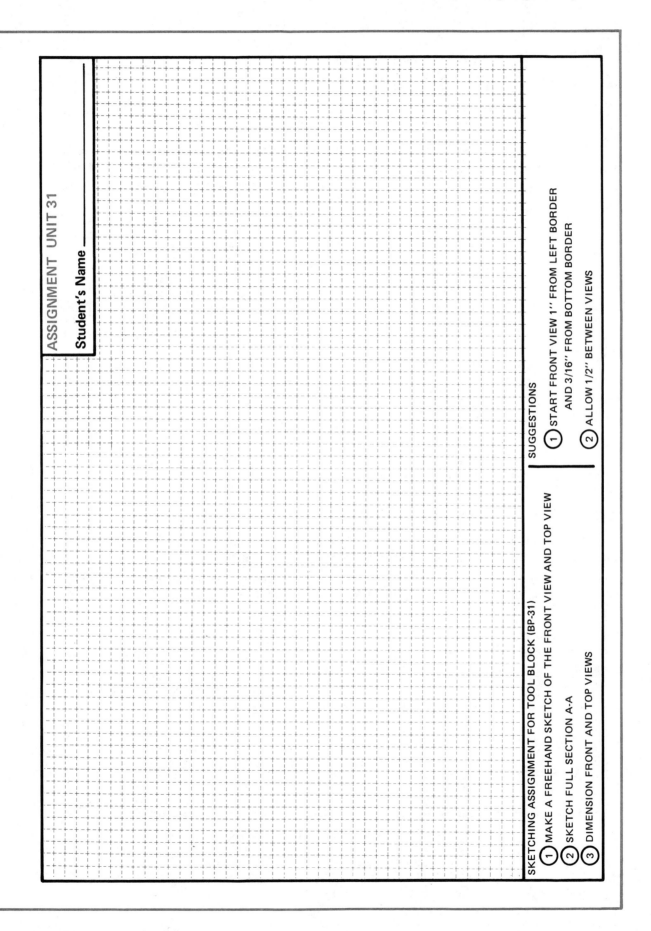

ASSIGNMENT UNIT 31

Student's Name _____

SKETCHING ASSIGNMENT FOR TOOL BLOCK (BP-31)

① MAKE A FREEHAND SKETCH OF THE FRONT VIEW AND TOP VIEW

② SKETCH FULL SECTION A-A

③ DIMENSION FRONT AND TOP VIEWS

SUGGESTIONS

① START FRONT VIEW 1'' FROM LEFT BORDER
   AND 3/16'' FROM BOTTOM BORDER

② ALLOW 1/2'' BETWEEN VIEWS

UNIT 32
Sketching Fillets, Radii, and Rounded Corners and Edges

Wherever practical, sharp corners are rounded and sharp edges are eliminated. This is true of fabricated and hardened parts where a sharp corner reduces the strength of the object. It is also true of castings where sharp edges and corners may be difficult to produce, figure 32-1.

FIGURE 32-1 Eliminating sharp edges and corners.

Where an outside edge is rounded, the convex edge is called a *round edge* or *radius*. A rounded inside corner is known as a *fillet*.

SKETCHING A FILLET OR RADIUS

Regardless of whether a fillet or radius is required, the steps for sketching each one are identical. The step-by-step procedure is illustrated in figure 32-2.

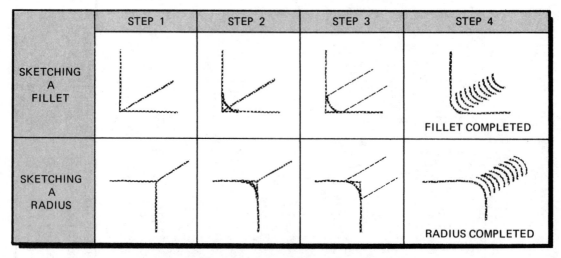

FIGURE 32-2 Steps in sketching a fillet or radius.

STEP 1 ▶ Sketch the lines which represent the edge or corner.

STEP 2 ▶ Draw the arc to the required radius.

STEP 3 ▶ Draw two lines parallel to the edge line from the two points where the arc touches the straight lines.

STEP 4 ▶ Sketch a series of curved lines across the work with the same arc as the object line. These curved lines start at one of the parallel lines and terminate at the other end.

SKETCHING INTERSECTION RADII AND FILLETS

The method of sketching corners at which radii or fillets come together is shown in the three steps in figure 32-3.

FIGURE 32-3 Sketching rounded corners and edges.

SKETCHING CORNERS ON CIRCULAR PARTS

The direction of radius or fillet lines changes on circular objects at a center line. The curved lines tend to straighten as they approach the center line and then slowly curve in the opposite direction beyond that point. The direction of curved lines for outside radii of round parts is shown in figure 32-4A. The curved lines for the intersecting corners are shown in figure 32-4B.

FIGURE 32-4 Direction of curved lines on round parts.

DROP FORGED STRAP | **BP-32A**

ASSIGNMENT A   UNIT 32

STUDENT'S NAME _____

SKETCHING ASSIGNMENT FOR DROP FORGED STRAP (BP-32A)
1) COMPLETE THE SKETCH BY SHADING THE FILLETED CORNER
     AND THE ROUNDED EDGES.
2) DIMENSION THE SKETCH. PLACE DIMENSIONS IN THE CIRCLES
     PROVIDED FOR THEM.

SHAFT SUPPORT | BP-32B

OIL HOLE
$\frac{1}{8}$ DRILL

4.00
2.00
2.50
1.62
.75

1.375 REAM
2.62
1.25
1.00
2.50

ASSIGNMENT B   UNIT 32

STUDENT'S NAME

SKETCHING ASSIGNMENT FOR SHAFT SUPPORT (BP-32B)

1 COMPLETE THE SKETCH AND SHADE THE ROUNDED CORNERS AND FILLET.

2 DIMENSION THE SKETCH. PLACE DIMENSIONS IN CIRCLES
        PROVIDED FOR THEM.

SECTION 9
Freehand Lettering

UNIT 33
Freehand Vertical Lettering

The true shape of a part or mechanism may be described accurately on a drawing by using combinations of lines and views. Added to these are the lettering, which supplies additional information, and the dimensions. For exceptionally accurate work and to standardize the shape and size of the letters, the lettering is done with a guide. In most other cases, the letters are formed freehand.

An almost square style letter known as *Gothic lettering* is very widely used because it is legible and the individual letters are simple enough to be made quickly and accurately. Gothic letters may be either vertical (straight) or inclined (slant), and upper or lower case.

FORMING UPPER CASE LETTERS

The shape of each vertical letter (both upper and lower case) and each number will be discussed in this unit. All of these letters and numbers are formed by combining vertical, horizontal, slant, and curved lines. The upper case letters should be started first as they are easiest to make.

Very light guide lines should be used to keep the letters straight and of uniform height. A soft pencil is recommended for lettering because it is possible to guide the pencil easily to form good letters.

LETTERS FORMED WITH
VERTICAL AND HORIZONTAL LINES

LETTERS FORMED WITH VERTICAL, HORIZONTAL AND SLANT LINES

LETTERS FORMED WITH STRAIGHT AND CURVED LINES

FIGURE 33-1 Forming upper case letters.

The shape of each vertical letter and number is given in figures 33-1, 33-2, and 33-3. The fine (light) lines with arrows which appear with every letter give the direction and number of the strokes needed to form the letter. The small numbers indicate the sequence of the strokes. Note that most letters are narrower than they are long. Lettering is usually done with a single stroke to keep the line weight of each letter uniform.

152

FIGURE 33-2 Forming vertical numerals.

SPACING LETTERS

To achieve good lettering, attention must be given to the proper spacing between letters, words and lines. Words and lines that are either condensed and run together or spread out are difficult to read, cause inaccuracies, and detract from an otherwise good drawing.

The space between letters should be about one-fourth the width of a regular letter. For example, the slant line of the letter A should be one-quarter letter width away from the top line of the letter T. Judgment must be used in the amount of white space left between letters so that it is as equal as possible. The letters will then look in balance and will be easy to read.

Between words, a space two-thirds the full width of a normal letter should be used. Lines of lettering are easiest to read when a space of from one-half to the full height of the letters is left between the lines.

LOWER CASE VERTICAL LETTERS

Lower case vertical letters are formed in a manner different from that used to form upper case letters. The main portion, or body, of most Gothic lower case letters is approximately the same width and height. The parts that extend above or below the body are one-half the height of the body. The shape of each vertical lower case letter and the forming of the letters are indicated in figure 33-3.

FIGURE 33-3 Forming vertical lower-case letters.

The spacing between lower case letters and words is the same as for capitals. The spacing between lines should be equal to the height of the body of the lower case letter to allow for the lines extending above and below the body of certain letters.

LETTERING FRACTIONS

Since fractions are important, they must not be subordinate to any of the lettering. The height of each number in the numerator and denominator must be at least two-thirds the height of a whole number. The dividing line of the fraction is in the center of the whole number. There must be a space between the numerator, the dividing line, and the denominator so that neither number touches the line, figure 33-4.

NUMERATOR AT LEAST TWO THIRDS
HEIGHT OF WHOLE NUMBER

IN CENTER OF WHOLE NUMBER

DENOMINATOR SAME HEIGHT
AS NUMERATOR

FIGURE 33-4 Size of parts of a fraction.

LETTERING RIGHT OR LEFT HANDED

The forming of letters to this point has been in terms of lettering with the right hand. Analysis of the steps followed in lettering with the left hand shows that the strokes are usually the same as for right hand. Occasionally, some of the lettering strokes are reversed. Figure 33-5 may be used as a guide for forming letters with the left hand.

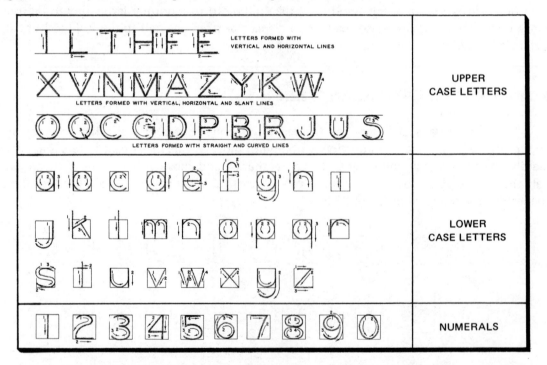

FIGURE 33-5 Forming vertical letters and numerals with the left hand.

The beginner should pay very close attention to the proper forming of letters, uniform line weights, and correct spacing. Once these skills have been mastered speed in lettering can be gained by continual practice in lettering words and sentences freehand.

① ASSIGNMENT UNIT 33

Student's Name _____

A B C D E F G H I

J K L M N O P Q R

S T U V W X Y Z &

2 3 4 5 6 7 8 9 0

a b c d e f g h i

j k l m n o p q r

s t u v w x y z

UNLESS OTHERWISE SPECIFIED
LIMITS ON DIMENSIONS ARE:
   FRACTIONAL DIMENSIONS $\pm\frac{1}{64}$"
   DECIMAL DIMENSIONS $\pm$.001"
   ANGULAR DIMENSIONS $\pm\frac{1}{2}$°

②

SKETCHING ASSIGNMENT FOR FREEHAND VERTICAL LETTERING (BP-33)

① LETTER FREEHAND THE UPPER AND LOWER CASE STRAIGHT LETTERS
     AND NUMERALS IN THE RULED SPACES PROVIDED

② LETTER THE NOTE ON TOLERANCES IN UPPER CASE LETTERS

UNIT 34
Freehand Inclined Lettering

Inclined letters and numbers are used on many drawings as they can be formed with a very natural movement and a slant similar to that used in everyday writing. The shape of the inclined or slant letter is the same as that of the vertical or straight letter except that circles and parts of circles are elliptical and the axis of each letter is at an angle, figure 34-1. Slant letters may be formed with the left hand using the same techniques of shaping and spacing as are used by the right hand. The only difference is that for some letters and numerals the strokes are reversed.

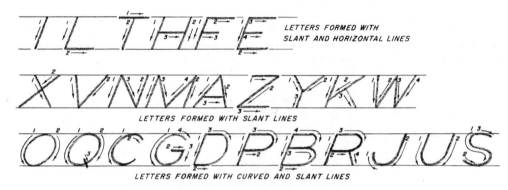

FIGURE 34-1 Forming upper case slant letters.

FORMING SLANT LETTERS AND NUMERALS

The angle of slant of the letters and numerals may vary, depending on individual preference. Lettering at an angle of from 60 to 75 degrees is practical, as it is easy to read and produce. Light horizontal guide lines to assure the uniform height of letters are recommended for the beginner. Angle guide lines assist the beginner in keeping all the letters shaped correctly and spaced properly. The direction of the strokes for each letter and numeral, and the spacing between letters, words, and lines are the same as for straight letters. The shape of each upper case slant letter is illustrated in figure 34-1; each numeral is shown in figure 34-2; and each lower case slant letter is shown in figure 34-3.

FIGURE 34-2 Forming slant numerals.

FIGURE 34-3 Forming lower case slant letters.

ASSIGNMENT UNIT 34

Student's Name _____

A   B   C   D   E   F   G   H   I

J   K   L   M   N   O   P   Q   R

S   T   U   V   W   X   Y   Z   8

2   3   4   5   6   7   8   9   O

a   b   c   d   e   f   g   h   i

j   k   l   m   n   o   p   q   r

s   t   u   v   w   x   y   z

UNLESS OTHERWISE SPECIFIED
LIMITS ON DIMENSIONS ARE:

FRACTIONAL DIMENSIONS $\pm \frac{1}{64}''$

DECIMAL DIMENSIONS $\pm .001''$

ANGULAR DIMENSIONS $\pm \frac{1}{2}°$

SKETCHING ASSIGNMENT FOR FREEHAND SLANT LETTERING (BP-34)

① LETTER FREEHAND THE UPPER AND LOWER CASE SLANT LETTERS
AND NUMERALS IN THE RULED SPACES PROVIDED

② LETTER THE NOTE ON TOLERANCES IN UPPER CASE LETTERS

SECTION 10
Shop Sketching: Pictorial Drawings

UNIT 35
Orthographic Sketching

PICTORIAL DRAWINGS

Pictorial drawings are very easy to understand because they show an object as it appears to the person viewing it. Pictorial drawings show the length, width, and height of an object in a single view. The use of a pictorial drawing enables an individual, inexperienced in interpreting drawings, to visualize quickly the shape of single parts or various components in a complicated mechanism.

FIGURE 35-1 Four common types of sketches.

| ORTHOGRAPHIC | OBLIQUE | ISOMETRIC | PERSPECTIVE |

There are three general types of pictorial drawings in common use: ① oblique, ② isometric and ③ perspective. A fourth type of freehand drawing is the orthographic sketch, figure 35-1. The advantages and general principles of making orthographic sketches are described in this unit.

| THREE-VIEW ORTHOGRAPHIC SKETCH | TWO-VIEW ORTHOGRAPHIC SKETCH |

FIGURE 35-2 Examples of orthographic sketches.

MAKING ORTHOGRAPHIC SKETCHES

The orthographic sketch is the simplest type to make of the four types of sketches. The views (figure 35-2) are developed as in any regular mechanical drawing and the same types of lines are used. The only difference is that orthographic sketches are drawn freehand. In actual practice, on-the-spot sketches of small parts are made as near the actual size as possible.

158

ROCKER FIXTURE   BP-35A

ASSIGNMENT A   UNIT 35

STUDENT'S NAME _____

SKETCHING ASSIGNMENT FOR ROCKER FIXTURE
① MAKE A FREEHAND SKETCH OF THE
   FRONT AND RIGHT SIDE VIEWS
② DIMENSION THE VIEWS

SUGGESTIONS
① DRAW TO HALF SIZE SCALE
   (EACH SQUARE = $\frac{1''}{4}$)

NO. REQ.	10
ORDER NO.	5 – 732
MAT'L.	CAST STEEL

GUIDE BRACKET | **BP-35B**

ASSIGNMENT B   UNIT 35

STUDENT'S NAME _____

SKETCHING ASSIGNMENT FOR GUIDE BRACKET (BP-35B)

① SKETCH FREEHAND THE REQUIRED AUXILIARY VIEW IN THE SPACE
   PROVIDED

② SKETCH FREEHAND AND DIMENSION FULL SECTION VIEW A-A

GUIDE BRACKET (BP-35B)

1. Name each of the three views.

2. Give the number and diameter of the drilled holes.

3. Determine dimensions Ⓐ and Ⓑ .

4. Compute dimension Ⓒ .

5. What is the dimension of the leg Ⓓ ?

6. Compute dimension Ⓔ .

7. Give dimension Ⓕ and Ⓖ .

8. What is the overall length of Ⓗ ?

9. Determine overall height of Ⓘ .

10. Give overall width of Ⓙ .

ASSIGNMENT B UNIT 35

Student's Name _____

1. _____ , _____ , _____

2. No. _____   Dia _____

3. Ⓐ = _____   Ⓑ = _____

4. Ⓒ = _____

5. Ⓓ = _____

6. Ⓔ = _____

7. Ⓕ = _____   Ⓖ = _____

8. Ⓗ = _____

9. Ⓘ = _____

10. Ⓙ = _____

UNIT 36
Oblique Sketching

An oblique sketch is a type of pictorial drawing on which two or more surfaces are shown at one time on one drawing. The front face of the object is sketched in the same manner as the front view of either an orthographic sketch or a mechanical drawing. All of the straight, inclined, and curved lines on the front plane of the object will appear in their true size and shape on this front face. Since the other sides of the object are sketched at an angle, the surfaces and lines are not shown in their true size and shape.

MAKING OBLIQUE SKETCHES WITH STRAIGHT LINES

The steps in making an oblique sketch are simple. For example, if an oblique sketch of a rectangular die block is needed, seven basic steps are followed as shown in figure 36-1.

STEP 1 ▶ Select one view of the object that gives most of the desired information.

STEP 2 ▶ Draw light horizontal and vertical base lines.

STEP 3 ▶ Lay out the edges of the die block in the front view from the base lines. All vertical lines on the front face of the object will be parallel to the vertical base line; all horizontal lines will be parallel to the horizontal base line. All lines will be in their true size and shape in this view.

FIGURE 36-1 Steps in making an oblique sketch.

STEP 4 ▶ Start at the intersection of the vertical and horizontal base lines and draw a line at an angle of 45° to the base line. This line is called the *oblique base line.*

STEP 5 ▶ Draw the remaining lines for the right side and top view parallel to either the oblique base line, or to the horizontal or vertical base lines, as the case may be. Since these lines are not in their true size or shape, they should be drawn so they appear in proportion to the front view.

STEP 6 ▶ Place dimensions so they are parallel to the axis lines.

STEP 7 ▶ Erase unnecessary lines. Darken object lines to make the sketch clearer and easier to interpret.

SKETCHING CIRCLES IN OBLIQUE

A circle or arc located on the front face of an oblique sketch is drawn in its true size and shape. However, since the top and side views are distorted, a circle or arc will be elliptical in these views. Three circles drawn in oblique on the top, right side, and front of a cube are shown in figure 36-2.

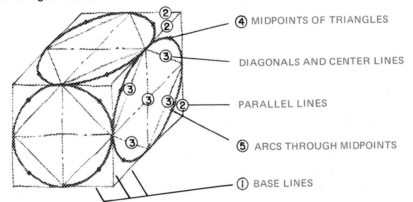

④ MIDPOINTS OF TRIANGLES

DIAGONALS AND CENTER LINES

PARALLEL LINES

⑤ ARCS THROUGH MIDPOINTS

① BASE LINES

FIGURE 36-2 Sketching circles in oblique.

STEP 1 ▶ Draw horizontal, vertical, and oblique base lines.

STEP 2 ▶ Sketch lines parallel to these base lines or *axes* to form a cube.

STEP 3 ▶ Draw center lines and diagonals in the front, right side, and top faces.

STEP 4 ▶ Locate the midpoints of triangles formed in the three faces.

STEP 5 ▶ Draw curved lines through these points. NOTE: The circle appears in its true size and shape in the front face, and as an ellipse in the right side and top faces.

STEP 6 ▶ Touch up and darken the curved lines. Erase guidelines where they are of no value in reading the sketch.

RIGHT AND LEFT OBLIQUE SKETCHES

Up to this point, the object has been viewed from the right side. In many cases, the left side of the object must be sketched because it contains better details. In these instances, the oblique base line or axis is 45° from the horizontal base line, starting from the left edge of the object. A rectangular die block sketched from the right side is shown in figure 36-3A. The same rectangular die block is shown in figure 36-3B as it would appear when sketched from the left side.

(A) RIGHT OBLIQUE   (B) LEFT OBLIQUE

FIGURE 36-3 Right and left oblique sketches.

The same principles and techniques of making oblique sketches apply, regardless of whether the object is drawn in the right or left position.

FORESHORTENING

When a line in the side and top views is drawn in the same proportion as lines are in the front view, the object may appear to be distorted and longer than it actually is. To correct the distortion, the lines are drawn shorter than actual size so the sketch of the part looks balanced. This drafting technique is called *foreshortening*.

In figure 36-4, the oblique sketch shows a part as it would look before and after foreshortening. The foreshortened version is preferred. The amount the sketch is foreshortened depends on the ability of the individual to make the sketch resemble the part as closely as possible.

FORESHORTENED

FIGURE 36-4 Foreshortening corrects distortion.

UNIT 37
Isometric Sketching

Isometric and oblique sketches are similar in that they are another form of pictorial drawing in which two or more surfaces may be illustrated in one view. The isometric sketch is built around three major lines called isometric base lines or axes, figure 37-1. The right side and left side isometric base lines each form an angle of 30° with the horizontal, and the vertical axis line forms an angle of 90° with the horizontal base line.

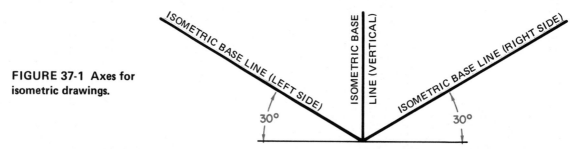

FIGURE 37-1 Axes for isometric drawings.

MAKING A SIMPLE ISOMETRIC SKETCH

An object is sketched in isometric by positioning it so that the part seems to rest on one corner. When making an isometric sketch of a part where all the surfaces or corners are parallel or at right angles to each other, the basic steps used are shown in figure 37-2. A rectangular steel block 1" thick, 2" wide, and 3" long is used as an example.

FIGURE 37-2 Steps in making an isometric sketch.

STEP 1 ▶ Sketch the three isometric axes. If a ruled isometric sheet is available, select three lines for the major axes.

STEP 2 ▶ Lay off the 3" length along the right axis, the 2" width on the left axis line, and the 1" height on the vertical axis line.

STEP 3 ▶ Draw lines from these layout points parallel to the three axes. Note that all parallel lines on the object are parallel on the sketch.

STEP 4 ▶ Dimension the sketch. On isometric sketches, the dimensions are placed parallel to the edges.

SKETCHING SLANT LINES IN ISOMETRIC

Only those lines that are parallel to the axes may be measured in their true lengths. Slant lines representing inclined surfaces are not shown in their true lengths in isometric sketches. In most cases, the slant lines for an object (such as the casting shown in figure 37-3) may be drawn by following the steps shown in figure 37-4.

FIGURE 37-3 Part to be sketched in isometric.

STEP 1 ▶ Draw the three major isometric axes.

STEP 2 ▶ Measure distance AB on the left axis and AC on the right axis.

STEP 3 ▶ Measure distance AD on the right axis and draw a vertical line from this point. Lay out distance DE on this line.

STEP 4 ▶ Draw parallel lines EF, BC', FD'.

STEP 5 ▶ Connect points A-E, B-F, and C-E.

STEP 6 ▶ Darken lines and dimension the isometric sketch.

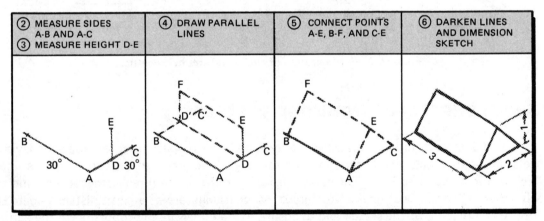

FIGURE 37-4 Isometric sketch requiring use of slant lines.

SKETCHING CIRCLES AND ARCS IN ISOMETRIC

The techniques used to sketch arcs and circles in oblique also may be used for sketching arcs and circles in isometric. Each step is illustrated in figure 37-5. Note that in each face the circle appears as an ellipse. This is also shown in figure 37-6.

FIGURE 37-5 Sketching circles in isometric.

FIGURE 37-6 Application of circles and arcs in isometric.

LAYOUT SHEETS FOR ISOMETRIC SKETCHES

The making of isometric drawings can be simplified if specially ruled isometric layout sheets are used. Considerable time may be saved as these graph sheets provide guide lines which run in three directions. The guide lines are parallel to the three isometric axes and to each other. The diagonal lines of the graph paper usually are at a given distance apart to simplify the making of the sketch and to ensure that all lines are in proportion to the actual size. The ruled lines are printed on a heavy paper which can be used over and over again as an underlay sheet for tracing paper.

CLUTCH GUIDE | BP-37

MATERIAL
CAST IRON
QUANTITY-25

$\frac{1}{8}$ R

ASSIGNMENT   UNIT 37

STUDENT'S NAME _____

SKETCHING ASSIGNMENT FOR CLUTCH GUIDE (BP-37)
1. MAKE A FREEHAND ISOMETRIC SKETCH OF THE CLUTCH GUIDE
2. DIMENSION THE SKETCH COMPLETELY

UNIT 38
Perspective Sketching

Perspective sketches show two or three sides of an object in one view. As a result, they resemble a photographic picture. In both the perspective sketch and the photograph, the portion of the object which is closest to the observer is the largest. The parts that are farthest away are smaller. Lines and surfaces on perspective sketches become smaller and come closer together as the distance from the eye increases. Eventually they seem to disappear at an imaginary horizon.

SINGLE POINT PARALLEL PERSPECTIVE

There are two types of perspective sketches commonly used in the shop. The first type is known as *single point* or *parallel perspective.* In this case, one face of the object in parallel perspective is sketched in its true size and shape, the same as in an orthographic sketch. For example, the edges of the front face of a cube as shown in parallel perspective in figure 38-1 at Ⓐ , Ⓑ , Ⓒ and Ⓓ , are parallel and square.

FIGURE 38-1 Single point, parallel perspective sketches of a cube.

To draw the remainder of the cube, the two sides decrease in size as they approach the horizon. On this imaginary horizon, which is supposed to be at eye level, the point where the lines come together is called the *vanishing point.* This vanishing point may be above or below the object, or to the right or left of it, figure 38-1. Vertical lines on the object are vertical on parallel perspective sketches and do not converge. Single point parallel perspective is the simplest type of perspective to understand and the easiest to sketch.

ANGULAR OR TWO-POINT PERSPECTIVE

The second type of perspective sketch is the two-point perspective, also known as *angular perspective.* As the name implies, two vanishing points on the horizon are used and all lines converge toward these points. Usually, in a freehand perspective sketch, the horizon is in a horizontal position. Seven basic steps are required to make a two-point perspective of a cube, using right and left vanishing points. The application of each of these steps is shown in figure 38-2.

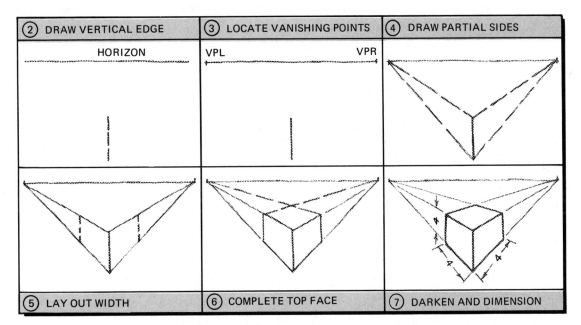

FIGURE 38-2 Making an angular perspective sketch of a cube.

STEP 1 ▶ Sketch a light horizontal line for the horizon. Then position the object above or below this line so the right vertical edge of the cube becomes the center of the sketch.

STEP 2 ▶ Draw a vertical line for this edge of the cube.

STEP 3 ▶ Place two vanishing points on the horizon: one to the right and the other to the left of the object.

STEP 4 ▶ Draw light lines from the corners of the vertical line to the vanishing points.

STEP 5 ▶ Lay out the width of the cube and draw parallel vertical lines.

STEP 6 ▶ Sketch the two remaining lines for the top, starting at the points where the two vertical lines intersect the top edge of the cube.

STEP 7 ▶ Darken all object lines and dimension. Once again, each dimension is placed parallel to the edge which it measures.

SKETCHING CIRCLES AND ARCS IN PERSPECTIVE

Circles and arcs are distorted in all views of perspective drawings except in one face of a parallel perspective drawing. Circles are drawn in perspective, using the same techniques of blocking-in which apply to orthographic, oblique, and isometric sketches. The steps in drawing circles in perspective are summarized in figure 38-3. The same practices may be applied to sketching arcs.

STEP 1 ▶ Lay out the four sides of the square which correspond to the diameter of the required circle. Note that two of the sides converge toward the vanishing point.

STEP 2 ▶ Draw the center lines.

STEP 3 ▶ Draw the diagonals.

STEP 4 ▶ Locate the midpoints of the triangles which are formed.

STEP 5 ▶ Sketch the circle through the points where the center lines touch the sides of the square and the midpoints.

⑤ SKETCH CIRCLE
④ LOCATE MIDPOINTS
③ DRAW DIAGONALS
② DRAW CENTER LINES
① COMPLETE SQUARE

FIGURE 38-3 Sketching circles in perspective.

SHADING PERSPECTIVE SKETCHES

Many perspective sketches are made of parts which include fillets, rounds, chamfers, and similar construction. By bringing out some of these details, the drawing becomes easier to read and is more attractive. The same techniques that are used for shading regular mechanical drawings or other pictorial sketches may be applied to perspective drawings. This shading brings out some of the construction details which otherwise might not be included, figure 38-4.

FIGURE 38-4 Shading applied to perspective sketch of die block.

SLIDE BLOCK

ASSIGNMENT   UNIT 38

STUDENT'S NAME _____

VPL   VPR

①

②

VPL

$\frac{3}{8}$DRILL, $\frac{9}{16}$ C'BORE, .25 DEEP, 2 HOLES

1.25

.50R

1.25

2.50

.50   2.75   2.75

6.50

1.00   1.00   1.25   1.25   1.00

.12R   .50R   .12R

.75   .62   2.50

SECTION
A–A

SKETCHING ASSIGNMENT ON PARALLEL PERSPECTIVE (BP-38)

① COMPLETE FREEHAND THE ANGULAR PERSPECTIVE SKETCH
     OF THE SLIDE BLOCK
② MAKE A FREEHAND PARALLEL PERSPECTIVE SKETCH OF
     SECTION A-A OF THE FIXTURE BASE. NOTE: USE THE
     GIVEN STARTING POINTS
③ SHOW CROSS-SECTION LINES, SHADE TO SHOW FILLETS, AND DIMENSION

UNIT 39
Pictorial Drawings
and Dimensions

PICTORIAL WORKING DRAWINGS

Many industries, engineering, design, sales, and other organizations use pictorial working drawings that are made freehand. In a great many cases, this method is preferred to instrument drawings. Technicians are also continuously making sketches in the shop and laboratory. The sketches simplify the reading of a drawing. The fact that no drawing instruments are required and sketches may be made quickly and on-the-spot makes freehand drawings practical.

A pictorial drawing is considered a working drawing when dimensions and other specifications that are needed to produce the part or assemble a mechanism are placed on the sketch. There are two general systems of dimensioning pictorial sketches: (1) *pictorial plane* (aligned) and (2) *unidirectional*.

PICTORIAL PLANE (ALIGNED) AND
UNIDIRECTIONAL DIMENSIONS

In the aligned or pictorial plane dimensioning system, the dimension lines, extension lines, and certain arrowheads are positioned parallel to the pictorial planes. These features, as they are positioned on a sketch, are illustrated in figure 39-1.

FIGURE 39-1 Pictorial plane (aligned) dimensioning.

By contrast, dimensions, notes, and technical details are lettered vertically in unidirectional dimensioning, figure 39-2. In this vertical position the dimensions are easier to read.

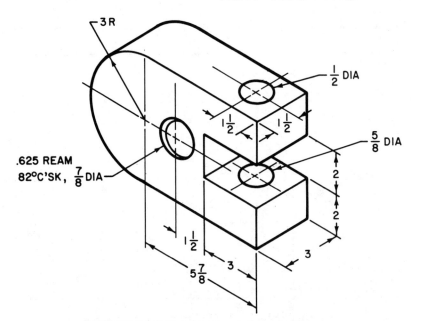

FIGURE 39-2 Unidirectional dimensioning.

Pictorial drawings are prepared according to both the United States and metric systems of representation. Thus, the dimensions, notes, and important data as included on the sketch, conform to the standards of the system being used.

PICTORIAL DIMENSIONING RULES

Essentially, the same basic rules for dimensioning a pictorial sketch are followed as for a multiview drawing. The basic pictorial dimensioning rules apply to both the aligned and unidirectional dimensioning systems.

RULE 1 ▶ Dimensions and extension lines are drawn parallel to the pictorial planes.

RULE 2 ▶ Dimensions are placed on visible features whenever possible.

RULE 3 ▶ Arrowheads lie in the same plane as extension and dimension lines.

RULE 4 ▶ Notes and dimensions are lettered parallel with the horizontal plane.

Since a pictorial drawing is a one-view drawing, it is not always possible to avoid dimensioning on the object, across dimension lines, or on hidden surfaces. These practices should be avoided in order to prevent errors in reading dimensions or interpreting particular features of the part.

Ø 34.925 $^{+0.04}_{-0.03}$

25.4 R

44.45

19.05

Ⓒ

Ⓓ 44.45 $^{+0.03}_{-0}$

5

73.1

Ⓑ

VIEW II

38.08 ±0.03

19.04 $^{+0.03}_{-0}$

Ø12.7 $^{+0.2}_{-0.1}$

11.1 R

Ⓐ

34.9

6.35

9.51 ±0.03

10

VIEW I

127

63.5

47.63

47.63

2 – Ø 5.95

9.9

Ⓔ Ⓕ

9.92 ±0.03

Ø 7
TWO HOLES

VIEW III

FIRST ∢ PROJECTION	UNSPECIFIED TOLERANCES +0.4 –0.2		
	DATE 4-20-82	PART SPECS 68-29 AST	NO. REQ. 12
DRAWN J.E.F.	CHKD C.G.W.	**BEARING HOUSING**	
C.G. OLIFOR MFG. CORPN.	MATL. COLD-DRAWN STEEL		BP-39

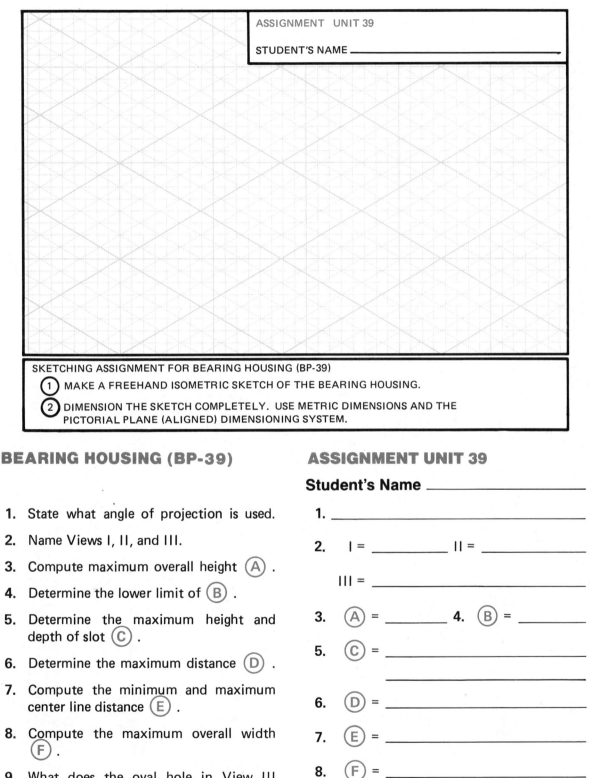

ASSIGNMENT   UNIT 39

STUDENT'S NAME _____

SKETCHING ASSIGNMENT FOR BEARING HOUSING (BP-39)

(1) MAKE A FREEHAND ISOMETRIC SKETCH OF THE BEARING HOUSING.

(2) DIMENSION THE SKETCH COMPLETELY. USE METRIC DIMENSIONS AND THE
PICTORIAL PLANE (ALIGNED) DIMENSIONING SYSTEM.

BEARING HOUSING (BP-39)

1. State what angle of projection is used.

2. Name Views I, II, and III.

3. Compute maximum overall height (A) .

4. Determine the lower limit of (B) .

5. Determine the maximum height and depth of slot (C) .

6. Determine the maximum distance (D) .

7. Compute the minimum and maximum center line distance (E) .

8. Compute the maximum overall width (F) .

9. What does the oval hole in View III represent?

10. Give the specifications for the counter-bored holes.

ASSIGNMENT UNIT 39

Student's Name _____

1. _____

2.   I = _____   II = _____

   III = _____

3. (A) = _____   4. (B) = _____

5. (C) = _____

   _____

6. (D) = _____

7. (E) = _____

8. (F) = _____

9. _____

10. _____

INDEX

A

Acme thread, 73
Adapter bracket, 115
Aligned method of dimensioning, 48–49
Alphabet of lines, 7
American National Standard thread, 73
American National Standards Institute (ANSI), 7, 72, 75
  standards
    drafting, 42
    for knurls, 88
  symbols
    for arc and gas welds, 117–118
    for standard welding, 119
    for surface finish, 84
American National threads, 72
American Standards Association, 72
American Welding Society, 119
Angle brace, 12
Angle bracket, 22, 23
Angles, dimensioning, 53
Angular dimensions, 65
  symbols used for, 65
Angular perspective, 171
Angular tolerances, specifying, 65
Aperture card, 4
Arc(s)
  connecting straight lines, 136
  connecting straight and curved lines, 136
  defined, 136
  dimensioning, 49
    with centers outside drawing, 60
  in perspective, 172
  sketching freehand, 136
Arc and gas welds, ANSI symbols for, 117–118
Arrowhead, 120
Auxiliary views, 38–41
  application of, 39
  primary, 39
  secondary, 39
Axes for isometric drawings, 166

B

Base line, defined, 60
Base line dimensioning, 60–61
  applications of, 61
Base plate, 62, 63
Bearing housing, 176, 177
Bilateral tolerances, 68
  equal/unequal, 105
Blocking, 144
Blue-line print, 5

Blueprint reading
  defined, 1
  as universal language, 1
Blueprinting, 2
Blueprints, bases for interpreting, 1–5
Break lines, 15
Breaks, conventional symbols for, 113
Broken-out section, 113
Brownprints, 3
Butt weld, 117
Buttress thread, 73

C

Cam plate, 10
Canadian Standards Association (CSA), 83
Center lines, 9
Centers, dimensioning, 56
Chamfers and grooves, dimensioning, 89
Change notes, 91
Circles
  dimensioning, 48
  invisible, 29
  in perspective, 172
  sketching, 137
    in isometric, 168
    in oblique, 163
Clutch guide, 169
Common Market, accepted standard of projection of, 101
Computer assisted machining (CAM) system, 124
Cone clutch, 114
Construction dimensions, 42–43
Continuous dimensions, 45
Controlling dimensions, 104
Coordinate charts, 125
Corner plate, 13
Corner weld, 117
Corners on circular parts, sketching, 149
Counterbored hole
  defined, 52
  dimensioning, 52–53
Countersunk hole
  defined, 53
  dimensioning, 53
Coupling, 58, 59
Cover plate, 34, 35
Cross-hatch lines, 108, 109
Cross slide, 24, 25
Curved lines, sketching, 136–143
Cutter adapter, 92, 93
Cutting plane(s), 108

Cutting plane line, 15, 108, 109
  and half section view, 112
Cylinders, dimensioning, 49

D

Datums, 124
Decimal dimensions, 68–69
Decimal tolerances, 64
  specifying, 69
Depth, 21
Depth of thread, 79
Diametral dimensions, 96
Diazo (Whiteprint) process, 5
  dry, 5
Die shoe, 46, 47
Dimension lines, 11
Dimensioning
  dual systems of, 104–105
  equally spaced holes, 56
  holes not on a circle, 57
  unequally spaced holes, 58
Dimensions, 21
  angular, 53
  continuous, 45
  in limited spaces, 45
  placing, 44–45
  size and location, 42–47
  standards for, 42
  two-place decimals, 43
  of weld, 120
Distortion, correcting, 164
Dovetailed slide, 70, 71
Drafting, language of, 42
Drawings, elements common to, 5
Drop forged strap, 150
Dryseal pressure-tight threads, 75
Dryseal straight pipe thread series, 75
Dryseal taper series, 75
Dual dimensioning, 105

E

Edge weld, 117
Electrostatic process, 3–4
End shaft, 30, 31
Extension lines, 11
  placing, 44
External screw threads, 74
  dimensioning, 75
  representation of, 74

F

Feeder plate, 16, 17
Field weld, 121
Fillet, 148
  sketching, 148–149

179

Finish All Over (FAO), 84
Finished surfaces, 83
First-angle drawings
    dimensioning, 101
    use of metric units in, 101
First-angle projection, 97, 101
    position of views, 101
Flange, 142
Flanged sleeve, 32, 33
Flash weld, 121
Flaws, in surfaces, 85
Foreshortening, 164
Forged rocker bracket, 106, 107
Forging, 83
    working drawing of, showing appli-
        cation of finish marks, 84
Fractional dimensions, limits on, 64
Fractional tolerances, 64
Freehand lettering, 152–157
    inclined, 156–157
Full section, 108
    view of interior, 109

G

Gear arm, 50, 51
Gothic lettering, 152
Graph paper, use of, 131
Groove angle, 121
Guide bracket, 160, 161

H

Half scale, 42
Half sections, 112
Heat process, 5
Height, 21
Hidden lines, 9
Holes
    dimensioning, 52
    equally spaced, 56
    not on circle, 57
    unequally spaced, 57
Horizontal lines, sketching, 130–131
Hub, 90

I

Ideas, conveying, 130
Idler shaft support, 110, 111
Inch system dimensions, conversion of, 94
Inclined lettering, freehand, 156–157
Industrial practices, and use of drawings, 2
Internal screw thread, 74
    representation of, 78
International Organization for Standard-
        ization (ISO), 96, 104
International Standards Organization
        (SI) Metric Thread System, 75
International System of Units (SI), 94
Intersection fillets, sketching, 149
Intersection radii, sketching, 149
Invisible circles, representing, 29
Invisible edge lines, 29
Irregular shapes, sketching, 144–147
Isometric sketching, 166–169
    circles and arcs in, 168
    layout sheets for, 168
    simple, basic steps in, 166
    slant lines in, 167

K

Key(s), 90
    specifications, 90
Keyseat, 90
Keyway, 90
Knurl(s)
    ANSI standards for, 88
    defined, 88
    pitch of, 89
Knurled surfaces, dimensioning, 88–89

L

Lap weld, 117
Lay, defined, 85
Layout sheets for isometric sketches, 168
Leader(s), 44
Left-hand threads, specifying, 79
Letter spacing, 153
Lettering
    with right or left hand, 154
    vertical, 152
Lettering fractions, 154
Linear dimensions, rounding off, 95
Lines, 7–17
    used in combination, 15
Location dimensions, 44
    for points, centers, and holes, 56–59
Lower case vertical letters, 153

M

Machined forging, 83
Mats, 4
Metric system, 94
    and inch system, converting, 94, 95
Microfilm, 4

N

National Pipe Taper (NPT), 75
National Screw Thread Commission, 72
Necking, 90
Nonsignificant zero, 105
Numerical control (NC)
    defined, 124
    functions, 124
    machine drawings, 124–125

O

Object lines, 7
Oblique sketching, 162–165
    circles in, 162
    making, with straight lines, 162–163
    right and left, 164
Offset carrier arm, 86, 87
One-view drawing, 36–37
    of cylindrical shaft, 36
    of flat part, 36
Ordinate dimensioning, 124–125
Orthographic sketching, 158–161
    making, 158

P

Partial section, 113
Part sizes. See Dimensions
Perspective sketching, 170–173
    angular or two-point, 171
    circles and arcs in, 172
    shading in, 172
    single point, parallel, 170

Phantom lines, 15
Photocopy, 4
Photostat, 4
Pictorial dimensioning rules, 175
Pictorial drawings, 158
Pictorial plane (aligned) dimensions,
        174–175
Pictorial representation of external
        screw threads, 74
Pictorial working drawings, 174
Pipe thread representation, 75
Points, dimensioning, 56
Positioning shaft, 98, 99
Positive positioning, 124
Pressure pad, 8
Projection, based on system of quadrants,
        100
Projection lines, 14
Projection symbols, 96–97
Pump system mount, 102, 103
Punch plate, 132

R

Radius, 148
    sketching, 148–149
Reference line(s), 120
    defined, 60
Reproduction
    defined, 2
    five basic processes, 2–5
Resistance welding, types and symbols,
        118–119
Right-hand threads, 79
Rocker base, 80, 81
Rocker fixture, 159
Root opening, 121
Rough forging, 82
Roughness, of surfaces, 85
Round edge, 148
Round groove, 89
Rounded corners and edges, sketching,
        149

S

Scale, 42
Schematic representation of external
        screw threads, 74
Screw threads, 72
    common 60° angle, 73
    conversion, 104
    dual systems of, 104–105
    external. See External screw threads
    left-handed, 78
    right-hand, 79
Section lines, 108, 109
    to identify materials, 109
Sectional view, 5
Sector punch, 66, 67
Sepia prints, 5
Series dimensions, placing, 45
Shaft, 138
Shaft support, 40, 41, 151
Sharp edges and corners, eliminating,
        148
Shop dimensions, 61
Shop notes, dimensioning with,
        88–91

Shop sketching. *See also* Sketches;
    Sketching; and *specific
    type*
  isometric, 166
  oblique, 162
  orthographic, 158
  perspective, 170
Shoulder pin, 37
Silver reproduction process, 4
Simplified representation of external
    screw threads, 74
Single point parallel perspective,
    170
Size dimensions, 43
  for holes and angles, 52–55
Sketches
  bases for interpreting, 1–5
  four common types of, 158
Sketching
  arcs in combination with other arcs,
    137
  curved lines and circles, 136–143
  freehand, 130
  horizontal lines, 130–131
  slant lines, 131
  value of, 130
  vertical lines, 131
Slant letters and numbers, forming,
    156
Slant lines, sketching, 131
Slide block, 140
Sliding support, 54, 55
Spindle shaft, 76, 77
Spot weld, 121
Square groove, 89
Square thread, 73
Stacker plate, 126, 127
Straight lines, types of used in drawing
    and sketching, 131
Straight pipe thread, 75
Surface finish, ANSI symbols for, 84
Surface texture
  defined, 85
  representation and measurement of,
    85

T
Tabular dimensioning, 125
Tap, defined, 78
Tap drill, 78
Tapers, 82
Tapped hole
  defined, 78
  dimensioning, 79
  dual dimensioning of, 104
Tapping, 78
Tee slide, 134
Tee weld, 117
Telecopier, 4
Third-angle system of projection,
    97
  principles of, 100
Thread class, 79
Thread form, 72
Thread series, 73–74
  coarse, 74
  constant pitch, 73
  extra fine, 74
  fine, 74
  nonstandard or special, 73
Three-view drawings, 18–25
  front, 18–19
  side, 20
  top, 19–20
Title block, 3
Tolerances, 64
  dual systems of, 104–105
  unilateral and bilateral, 68
Tool block, 146
Two-place decimal dimensions, 43
Two-view drawings, 28–35
  combination of views for,
    28–29

U
Undercutting, 90
Unidirectional dimensions, 174–175
  methods, 48–49
Unified Inch System, 104
Unified Metric, 104
Unified screw thread, 73

Unified (UN) thread, 72
  profile of, 72
Unified Thread System, 72
Unilateral tolerance, 68, 105
U.S. Standards Institute, 72
U.S. Standard threads, 72
Upper case letters, 152

V
"V" groove, 89
Vandyke, 5
Vanishing point, 170
Vertical lettering, freehand, 152
Vertical lines, sketching, 131
View(s), 5, 18–41
  abbreviation of, 26
  arrangement of, 26–27
  identification of, 26
  with transparent planes removed, 21
Visible edge. *See* Object lines

W
Waviness, in surfaces, 85
Weld
  contour of, 121
  finish of, 121
Weld symbol(s)
  basic, 120
  for resistance welds, 119
  supplementary, 119
Welded drawings, 117–120
Welded joints, types of, 117
Welded rotary plug, 122, 123
Welded structures and symbols, 117
Whiteprints, 5
  wet, 5
Width, 21
  and length, 21
Woodruff key, 90
Working drawings, 21
Workpiece program, 124

Z
Zero(s), nonsignificant, 105
Zero coordinates, 124